A message from our Director

Welcon

Welcon

Welcome to 2020

Welcome to 2022

Welcome to 2024

Welcome to 2025

Welcome to 2026

When people first learn about Institute for the Future, they generally have a lot of questions. The first is mostly, "What do you do?" (We're a non-profit think tank dedicated to helping people think through future possibilities so they can make better decisions today). The second is usually something along the lines of, "Do you predict the future?" (No, we don't. We create plausible internally consistent forecasts of what the world could be like 10 years from now.) The third question is often, "How large is IFTF?" It seems like the answer ought to be straightforward, but it's not. While our staff roster floats around the relatively small number of 40 or so people, in truth, IFTF is much bigger. To get robust perspectives on the future, we work with a globally distributed network of technologists, designers, journalists, activists, academics, science fiction writers, policy makers, and entrepreneurs in business and civic society. We consider this growing network of thinkers and doers an important part of our organization, and this publication, *Future Now*, is a reflection of that.

In this second volume of *Future Now,* we explore the future of communications. In our research process, we traced historical technology shifts through the present and focused on the question, "what is beyond social media?". Our journey started with an expert workshop in Silicon Valley, where we engaged some of the area's most creative and insightful thinkers to help us identify a set of foundational technologies shaping the next decade of communications. We then went to Copenhagen, where we convened a broad set of experts, speculative designers, and future thinkers to map out how people would use these technologies to fulfill their long-standing needs, desires, and intentions. Next, we traveled to New York City and met with another set of experts, including professionals in media and advertising, artists and students, and civic technology innovators and activists, to synthesize different possibilities into coherent stories about the future of communication.

It was through these conversations that we were able to uncover the emerging world of ambient communications, and began to identify the many new layers of complexity that will require us to rethink what's possible and reinvent what's practical in this rapidly approaching future.

The publication you're holding reflects the conversations and insights we've had over this past year about the future of technology and communications. Think of *Future Now* as a book of provocations; it reflects the curiosity and diversity of futures thinking across IFTF and its network of collaborators. It contains expert interviews, profiles and analyses of what today's technologies tell us about the next decade, as well as comics and science fiction stories that help us imagine what 2026 (and beyond) might look and feel like. You may find some of these provocations challenging, scary, or even ridiculous. The point is not to persuade you to agree with any particular forecast or point of view, but to provoke you to engage in more conversations about the future, and to consider new possibilities.

Welcome to the future, now.

Rod Falcon
Tech Horizons Program Director

Contents

FUTURE NOW

Technology Horizons Program Director
Rod Falcon

Senior Editor
Mark Frauenfelder

Research Director
Bradley Kreit

Managing Editor
Ben Hamamoto

Technology Horizons Team and IFTF Contributors
Miriam Lueck Avery, Jamais Cascio, Peter Coughlin, Devin Fidler, Alex Goldman, Eri Gentry, Marina Gorbis, Dylan Hendricks, Toshi Hoo, Bob Johansen, Lyn Jeffery, Chris Kalaboukis, Mike Liebhold, Jane McGonigal, Scott Minneman, David Pescovitz, Jason Tester, David Thigpen, Andrew Trabulsi, Kathi Vian, Anthony Weeks

Contributing Editors
Cindy Baskin, Gareth Branwyn, Andrew Keller, Carla Sinclair, Sara Skvirsky

Copy Editor
Nicola Morrison

Executive Producer
Jean Hagan

Program Manager
Meagan Jensen

Design
Albertson Design

Algo Character Design
John Hersey

Production Director
Robin Weiss

Production team
Robin Bogott, Maureen Kirchner, Karin Lubeck, Jeanne Schreiber

Illustration
Trent Kuhn, Mitch O'Connell, Hisashi Okawa, Webuyyourkids

Business Development
Dawn Alva, Daria Lamb, Neela Lazkani, Sean Ness

Printed in Canada

SR-1895A

Contributors

Madeline Ashby is a science-fiction writer and futurist living in Toronto. She has a Masters of Design in Strategic Foresight and Innovation from OCADU, and has worked with IFTF, SciFutures, Intel Labs, Nesta, Data & Society, and others. Her latest novel, "Company Town" is available from Tor Books.

Sam Gregory is a human-rights activist and founder of Witness, whose mission is to enable anyone to be an active witness for human rights using contemporary technologies.

Kevin Kelly is Senior Maverick at Wired magazine. He co-founded Wired in 1993, and served as its Executive Editor for its first seven years.

Claire Rowland is a London-based independent product/UX strategy consultant specialising in connected products/IoT. She is the lead author of "Designing Connected Products: UX for the Consumer Internet of Things," published by O'Reilly.

Rudy Rucker is a writer, a mathematician, and a painter. For 20 years he was a professor of computer science at San Jose State in Silicon Valley. He received Philip K. Dick awards for his cyberpunk novels "Software" and "Wetware," and an Emperor Norton award for his autobiography "Nested Scrolls". Recent novels include "Turing & Burroughs" and "The Big Aha".

Doug Rushkoff is a writer, documentarian, and lecturer whose work focuses on human autonomy in a digital age.

Carla Sinclair is an author, the co-founder of Boing Boing, and editor of Wink Books.

Experts consulted

David Birnbaum, danah boyd, Ming-Li Chai, Christian Cherene, Brian Christian, Angèle Christin, Mattia Crespi, Hannah Chung, Roger Davis, Lars Ebert, Devin Fidler, Noah Goodman, Michael Joaquin Grey, Dehlia Hannah, Anders Sahl Hansen, John Harbort, Justin Hendrix, Alan Henry, Laurin Herr, Aaron Horowitz, Fu-Chung Huang, Anders Hvid, Louise Opprud Jakobsen, Michael Kleeman, James Kotecki, Henrik Kristensen, Michael MacKay, Thomas Madsen-Mygdal, Lauren McCarthy, An Xiao Mina, Robert Morris, Mor Naaman, Eric Paulos, Emmanuel Quartey, Jan Rabbaey, Roope Raisamo, Geir Terje Ruud, Wendy Schultz, Daniel Stoller, Jacob Strand, Thomas Vestskov Terney, Mark Tocher, Gary Wolf, Adrian Zaugg

Why We Communicate

Intentions in a world of ambient communications

by Rod Falcon, Bradley Kreit, Ben Hamamoto, Jason Tester, Mark Frauenfelder Illustrations by Jacob Glaser

We are on the verge of a set of technological advances that will transform when, where, how, and why we communicate. The Internet of Things, advances in network speed, and emerging immersive media platforms are transforming our communications experiences. Communications infrastructure will be embedded into our physical world, with messaging and interactions that adapt to contextual needs and anticipate future moves in order to grab our attention. Flat visual interfaces will give way to experiences that engage our whole bodies in multisensory ways, illuminate invisible patterns, and enable us to program our communications streams and orchestrate massive computational resources with just a few casually spoken words.

This communications infrastructure will seamlessly merge our digital and physical worlds to add layers of meaning and responsiveness to human-to-human, machine-to-human, and machine-to-machine interactions. The next decade will usher in a world of ambient communications that will challenge us to radically rethink our communications strategies.

Familiar Intentions, New Experiences

To unlock the value of these emerging technological capacities, we need to understand how to connect them to familiar human intentions.

In this issue of Institute for the Future's *Future Now*, we explore how these technological capacities will intersect with long-standing intentions of communication through a series of scenarios, articles, interviews and other provocations. Collectively, they explore issues like how powerful new technologies of surveillance can be used for everything from making games more fun and engaging to maintaining intimate relationships between family members. We look at how sensors, artificial intelligence, blockchain, and blended reality tech will transform how we collaborate, allowing massive crowds to work together in emergent ways, and even allowing us to partner with animals and digital entities in new ways. As you explore the pieces organized across this range of eight intentions—fun, persuasion, collaboration, productivity, control, engagement, empathy, intimacy—look for specific strategies and combinations of technologies that can best serve an intention.

Visions of Ambient Communications

Shifts in our communications technologies accompany shifts in our typical, everyday practices. In just the past decade, now-familiar images of teenagers becoming celebrities in their bedrooms, street protesters reaching global audiences, and mobile knowledge workers turning any place into office space have gone from being difficult to imagine to so familiar they're almost mundane.

In the pages that directly follow, you'll find comics set in the year 2026 that will immerse you in this soon-to-be familiar landscape. From popstars using digital holograms as brand ambassadors to intelligent furniture that bonds families across distances, these scenes point to the kinds of communication experiences that will stretch your sense of possibility but will seem unremarkable a decade from now.

Technologies That Matter

New capacities of ambient communication experiences

Looking out to the year 2026, we see a ubiquitous, all-encompassing communications infrastructure that seamlessly merges our digital and physical worlds. This infrastructure combines configurable wireless networks, embedded sensors, cloud computing, and artificial intelligence that adds layers of meaning and responsiveness to human-to-human, machine-to-human, and machine-to-machine interactions.

These and other foundational technologies are combining in novel ways to create five new capacities of ambient communications that provide the infrastructure of the media experiences we will create to fulfill our intentions.

Embedded From extrinsic to intrinsic

Computation is a three-part act: accept an input signal, process the signal, and output the result. Computation has gotten cheaper, faster, and smaller over the decades. We've gone from multi-ton, multimillion-dollar mainframes in the mid-20th-century to $300 wristwatches embedded with a microphone, altimeter, accelerometer, gyroscope, optical pulse sensor, ambient light sensor, touch sensor, microelectromechanical system actuator, gigabytes of storage, and wireless Internet. Now imagine the functions of a smartwatch embedded into everything from eyewear to clothing to cereal boxes, eventually even into our bodies. In the coming years, the combined trends of Moore's Law and sensor fusion will enable the proliferation of tiny super machines connected by configurable mesh networks, embedded with artificial intelligence, sensors, and local data sets. These will generate and summarize metadata, perform translation and local analytics, and enable an immersive communications experience that's ubiquitous, hyperaware, and invisible.

Illuminated From heuristic to optimal

Yesterday's home thermostats used timers as non-optimal best guesses to control room temperature. Today's smart thermostats learn about the changing behavior of a building's occupants to establish an optimal temperature schedule. The next decade will see the rise of smart things that collect and analyze multiple streams of data to unveil previously hidden troves of understanding. Improvements in pattern recognition, computer vision, and artificial intelligence technology will shine a light on the dark matter of our individual and group behavior. AIs and robots will learn from us—and from each other—and will begin telling us new things about our homes, bodies, cities, and personalities. We'll tune into any combination of many layers of pervasive augmented realities, allowing us to see formerly invisible patterns, which will help us—and our automated proxies—make better decisions about our work schedules, travel plans, health regimens, and financial investments.

Anticipatory From reactive to proactive

Deeply linked information streams and improved machine intelligence are giving computational systems the ability to anticipate our needs and deliver advice and information in a timely and context-appropriate manner. For instance, applications such as Google Now scan our email for package deliveries, appointments, and interests to create highly personalized reports of actionable information. In the next decade, everything visible and tangible in immersive experiences will be activated, enriched, and supported by machine intelligence in ambient communications networks. Networks will make use of advanced deep learning to combine and analyze unstructured big data sets and construct computer-based models of the world. They will spot patterns within large data sets and infer what action may occur or be most effective. Such networks have the potential to craft messaging specifically to individuals or to identify trends even before they achieve conscious awareness among humans.

Multisensory From visual to embodied

For decades, computational output was limited to visual data—blinking lights, words punched onto teletype paper, green phosphor shapes floating across screens. Graphics grew more sophisticated and were joined by sounds. Today, we experience synthetic realities at resolutions rivaling the real world. Until recently, this high-fidelity communication was limited to outputs—people could input information only by entering machine-readable characters one at a time. Now computers are acquiring sensory organs, making natural forms of input possible. Siri and Alexa are ears connected to a natural-language system that hears what we need and delivers it digitally or orders it shipped. Google Glass and other smart eyewear devices see things we're looking at, including things we don't notice, and provide contextual information. Ambient-communications networks will engage our entire body by becoming more responsive to voices, gestures, emotional states, skin, and eventually thoughts. With holographic displays and haptic interfaces, our media won't merely connect with us, they will envelop us.

Programmable From solo to symphony

On their own, each of these emerging technology capacities will advance ambient-communications experiences. When orchestrated collaboratively around the world, using distributed automated workflows and high-speed fiber interconnection, and mediated by intelligent agents, they'll effect a tremendous social transformation. Interfaces will customize themselves for the user through machine learning. Blockchain technology will become integrated into devices and networks to conduct transactions and generate records, allowing us to autonomously transact on a peer-to-peer basis with every entity from friends, doctors, and financial advisors to trusted AIs. Every person, thing, place, idea, and media object will have one or more intelligent layers of data. Configurable device protocols will connect these layers through a common language, enabling new information systems to flourish. We'll enlist an array of machine-based systems to carry out our intentions. Because of their varied utility, ambient-communications networks will be adopted even more rapidly than preceding information technologies.

The SHOPPER and the BRAND

Step into a world where business-to-consumer communications gets personal. Here, shoppers get personalized advice from a humanoid chatbot in hologram form. In fact, they receive counsel from *multiple* bots, each representing a different desire the shopper has, from frugality to well-being, and from environmental responsibility to novelty. In this near-future world, shopbots are developed and deployed by

influential people, publications, and brands, providing recommendations in exchange for a cut of each sale. The more you buy, the more their owners make. But this creates a dilemma—bot owners can use data-driven insights to push the products consumers are most likely to purchase, but if they stray too far from their brand identity, they might undermine the trust that made them an authority in the first place.

The WORKER and the OUTLAW

Ever felt like you're just another cog in the machine? Well, in this world you're a subroutine. Here, workers receives task requests at all times of the day and communicate through a platform. An "employee" might never see or speak directly to their "boss." And for project managers (both legit and criminal) getting things done means building "recipes" that string together platforms, tasks, and people. When one

worker completes a task, it triggers a new worker to start the next task in the recipe. These recipes ensure everyone does their part in a particular sequence, resulting in a coherent outcome, without individual contributors ever needing to see the whole. The next time you accept a task in the platform economy, you may not know which cog, er, person you're collaborating with—or what you're an accomplice to.

The FAMILY and the NOMAD

Today, a live-in assistant is too costly for most families, at least in the United States. But what if that assistant was an app? Welcome to a world where intelligent digital agents like Siri are embedded into furniture and kitchen counters, interacting with the entire household. These family assistants do more than recite recipes or driving directions. They play a key role in keeping the family connected, mediating

conflicts, and coordinating schedules for activities outside the home. In this future, work and learning are much more improvisational than they were in 2016. Fewer events and opportunities occur at fixed times and locations. Everything is ad hoc and on-the-go. As always, the younger generation is creating new approaches to navigating space and time. The rest of us will have to learn to keep up.

The ACTIVIST and the POLITICIAN

No matter what you're interested in, there's a social-media personality you can turn to for information and entertainment. Activists can learn a lot from livestreaming stars when it comes to engaging and growing audiences around the issues they care about. In this future, activism is as much about presenting evidence as it is telling compelling stories. And sensors and software that collect and analyze

environmental and behavioral data will play a key role in creating stories that move us.

Politicians, meanwhile, get unprecedented access to data about their constituents and their needs. But the data isn't reported as numbers on a spreadsheet or words on a page—it's used to generate complex simulations that model the results of their policy proposals.

The ATHLETE and the REFUGEE

Front-row seats are the most coveted vantage point to view a sporting event. But what if you could experience an event from the athlete's point of view? In this future, drones capture the action from the best angle, and sensors record smells, sounds, and sensations, allowing anyone to share the athlete's experience with unprecedented intimacy.

But capturing an immersive experience isn't always about fun. Journalist Ghaith Abdul-Ahad remarked that refugees often perform feats of extreme athleticism on their journeys. In this world, ambient communications capture technology is deployed to give us new multisensory views into unfamiliar experiences, and to challenge stereotypical concepts about refugees by allowing us to share their journey.

My future—let me show you it

Hi. I'm Algo

"We're all immigrants to the future; none of us is a native in that land." At least, that's what Marina Gorbis, our executive director here at Institute for the Future says, paraphrasing cultural anthropologist Margaret Mead. Marina thinks we should look at the future through the eyes of newbies—ready to explore, to learn, to do things differently.

Whenever you enter unfamiliar territory, it helps to have a guidebook, and a guide. Your guidebook to the world of 2026 is this volume of *Future Now*. Your guide is me.

"Wait, who are you, again?" you might be asking. Sorry, should've introduced myself. I'm Algo, a bot created from the collective intelligence of the Tech Horizons' research staff.

You'll see me pop up throughout this volume of *Future Now* to offer additional information, share an insight, or provide more context. You'll find me popping up from time-to-time in the margins of articles, and at the end of each section to highlight the big takeaways. For now, my guidance is geared for a broad audience. And, unfortunately, I can't hear you when you ask me a question. But one day soon, I'll know a lot more about you, your interests and your knowledge gaps. I'll be everywhere you are, ready to have a conversation with you anytime you need me. That's because I'm a native of the world you're about to enter, immigrant. Welcome to the future of ambient communications!

Fu

Fun

INTRODUCTION

How will we provoke imagination and enjoy ourselves?

From educational toys implanted with smarts to e-sporting events and alternate-reality games with millions of ardent fans, the rise of ambient communications will amplify the intense, serious, and frivolous ways we already have fun, and create entirely new ways to be playful.

The definitive parents' guide to the hot new toys of 2026

Fu Fun

by Dylan Hendricks

Rock or Block?

As we wade deeper into the 2020s, the lines between fun, learning, and data-mining continue to blur beyond recognition. Every day it seems like there's a new toy on the market that's more platform than plaything, offering to transform your unsuspecting offspring's way of life in exchange for a monthly subscription fee and unfettered access to their moldable little minds. Our intrepid research staff have opened their hearts and wallets to this year's ecosystem of super-powered toys, sorting the prime virtual wheat from the 3D-printed chaff. What new toys will rock your children's world, and which ones should you kick out of the sandbox? We're here to help.

Best New Toy to Replace Your Kid's Friends
HOBBS™ Virtual Buddy
Augmented Empire
Cost: $199 (plus subscription)

Like the wise-cracking animal sidekicks from every worthwhile children's movie, the HOBBS™ line of augmented plush dolls has always straddled the line between comforting ally and powerful virtual assistant. This year, Rhona Rhino and Paulie Panda return from extinction to join Augmented Empire's lineup of rugged, machine-washable toys, all of which spring to life in your favorite blended reality setup. With over a dozen algorithmic personalities available across 43 languages, HOBBS™ virtual buddies adapt to your child's behavior and learning styles, playing the role of invisible friend and furry mentor from childhood through early adolescence.

Fun — **Function**

HOBBS™ has many education-themed activities in its wheelhouse, but its strengths lie more with sass than skill sets.

Data Rich — **Data Risk**

HOBBS™ evolves alongside your child's developmental milestones, but doesn't share or sell their personal info in the process.

Solo — **Symphony**

While the virtual buddies all pal around together perfectly, try to pair them up with anything outside of Augmented Empire's toy roster and you're on your own.

Rock or Block? ROCK. Just because the love is software-based doesn't mean it isn't real.

Best New Toy to Replace Your Kid's Pet
Micro-Biodome Bacteria Farm
BioTots
Cost: $179 (plus subscription)

Now your young microbiologist-in-training can explore the wild zoology of the bacterial world with this surprisingly accessible bacteria farm from BioTots. Styled more like an upside-down fishbowl than the ant farm for which it's named, the Micro-Biodome can identify more than ten thousand different bacterial varieties within a 5-inch pocket of air. From your living room carpet to the front lawn, the playful visualizations of each unique bacterial cloud offer hours of childhood wonder for scientists of all ages.

Fun — **Function**

Come for the colorful and imaginative circus, stay for the up-to-date information on your local bacterial soup.

Data Rich — **Data Risk**

Your Micro-Biodome's findings contribute directly to global bacterial research. This is one category of data that you don't want to keep to yourself.

Solo — **Symphony**

This microbiome garden is anything but walled. You'll be able to see its plush textures and vivid colors on all the surfaces and screens of your home, no matter what operating system they use.

Rock or Block? ROCK. There's nothing so raucous as a pet Micrococcus.

Worst New Toy to Replace Your Kid's Diaper
Poop-4-Me Potty Training Smart Toilet
BioTots
Cost: $239 (plus subscription)

Also from BioTots, the Poop-4-Me smart toilet wants to guide your little ones to diaper-free independence, but it's not the kids who will be pissing their money down the drain with this one. Leveraging the same biotech found in the MicroBiodome, the Poop-4-Me can analyze your child's daily deposits for infections and dietary irregularities. Unfortunately, these teachable moments aren't worth the hours you'll spend scrubbing misfired guano from the unit's many nooks. Between the condescending voice assistant and the inexplicably steep monthly subscription, you're better off cutting the bottom out of your toddler's coveralls.

Fun **Function**

More potty than pal, not that it excels at either.

Data Rich **Data Risk**

The toilet itself is pretty smart, but you'd still be stupid to drop over $200 on it.

Solo **Symphony**

You locked into paying the steep monthly subscription fees, so if you wanna take your business elsewhere, you'll be left scrambling for a place to drop it.

Rock or Block? BLOCK. Like its raison d'etre, this thing is a piece of crap.

Best New Toy to Replace Your Kid's Teacher/Babysitter/Guardian
BoosterPak Complete Learning System
Shahrazad Group Holding LLC
Cost: $277 (plus subscription)

Hugely popular in China, the BoosterPak Complete Learning System comes west this year, offering a unique combination of learning, gaming, and ambient childcare. The BoosterPak system consists of an all-in-one backpack and blended reality visor that will turn any physical environment into an immersive classroom and virtual playground. The Beijing-based curriculum operates at a frenetic and sometimes troubling pace, but Shahrazad promises integration with major American e-learning platforms in the near future. In the meantime, your child won't run out of things to do with the system anytime soon.

Fun **Function**

The BoosterPak favors curriculum over creative chaos, nudging your child to complete active missions before moving on to the next activity.

Data Rich **Data Risk**

We're pretty sure the BoosterPak leaks its activity data all over the known world, but doesn't require full data access to be usable.

Solo **Symphony**

Shahrazad is selective about its partners, which keeps the quality well under control. But if you're looking for harmony with your local school curriculum, you'll have to wait.

Rock or Block? ROCK. The BoosterPak is the wearable babysitter you didn't know you needed.

Best New Toy to Replace Your Kid's Dreams with Nightmares

Spammy Bear

Original manufacturer unknown

Cost: N/A

Everybody knows the story: a cheaply made stuffed animal shows up in your child's bedroom one day. It's emblazoned with the logos of obscure companies. It smells bad. It's ugly. It randomly blurts out corporate slogans. You try to throw it out, but a week later it's somehow back. Nobody knows who first manufactured the spammy bears, but teardowns have revealed that these unwanted teddy bots are hiding enough computing power to host a generic DAC (decentralized autonomous corporation). Programmed to infest, they can order on-demand gig workers to ferry themselves out of the dumpster and reemerge into your child's life.

Fun — **Function**

Freakshow. Failure. Fungus. Foe.

Data Rich — **Data Risk**

This particularly ambitious specimen carries always-on cameras and motion sensors, elevating the already terrifying risk of exposure.

Solo — **Symphony**

Spammy Bear will pair with anything and everything to keep itself in your life—including holding your bank account, biodata and personal correspondence hostage.

Rock or Block? BLOCK. Like a bad horror movie, they won't stop until you cut off their heads and burn the stuffing.

Best New Toy to Replace Your Kid's Toys (Special Mention for Resource-Strapped Families)

Malaika Super Doll

AfriCare Communication Systems

Cost: $35–$79 (plus subscription)

Named for benevolent spirits from East African folklore, the Malaika Super Doll is more than just a toy, it's an investment. Beneath the Makonde-styled exterior, these dolls are mostly battery, offering a generous 50000mAh of juice for supercharging your family's lifestyle (or just recharging your devices). Already a mainstay from Sudan to Mozambique, these ornate figurines have gained popularity in recent years with hip urban nomads and international migrant camps. When the grid's unreliable, this is one mythological spirit whose generosity won't go unnoticed.

Fun — **Function**

Admittedly, the Super Doll doesn't really *do* much, but maybe you can convince your kid to lug it around for the good of the family?

Data Rich — **Data Risk**

This is one of the few toys left on the market that requires no data at all, and gives much more than it takes.

Solo — **Symphony**

Malaika's base is stuffed with sockets in every different size, shape, and format you might need to charge any device you could conceivably have sitting around.

Rock or Block? ROCK. Malaika will warm your child's heart, so why not let it warm your hearth?

The surprising ways augmented reality games affect our brains and bodies

Fu Fun

by Jane McGonigal

Power Up and Evolve: It's Not Just for "Pokémon"

In early July, "Pokémon Go" burst onto the smartphone screens of people around the globe seemingly overnight. Within the first week of its release, the game had more daily users than Twitter and more installs than Tinder or Snapchat. Within three weeks, 75 million people around the world were playing it every day—making it not only the fastest growing app in the world, but the fastest growing product in human history.

In terms of sales and cultural impact, "Pokémon Go" is an undeniable coup for its creators—and for context-aware, augmented-reality gaming in general. But there's a more important way to measure its success: in human well-being.

A staggering third of all players reportedly count weight loss among their motivations for playing—and data collected by major news outlets indicate they have been wildly successful. I crunched the numbers with my math friends on Twitter and we estimated that people playing "Pokémon Go" were collectively losing 571,000 pounds a day. And that's not all. While it's harder to quantify, major mental health benefits are being reported as well—particularly for kids with autism and people who suffer from chronic social anxiety and depression.

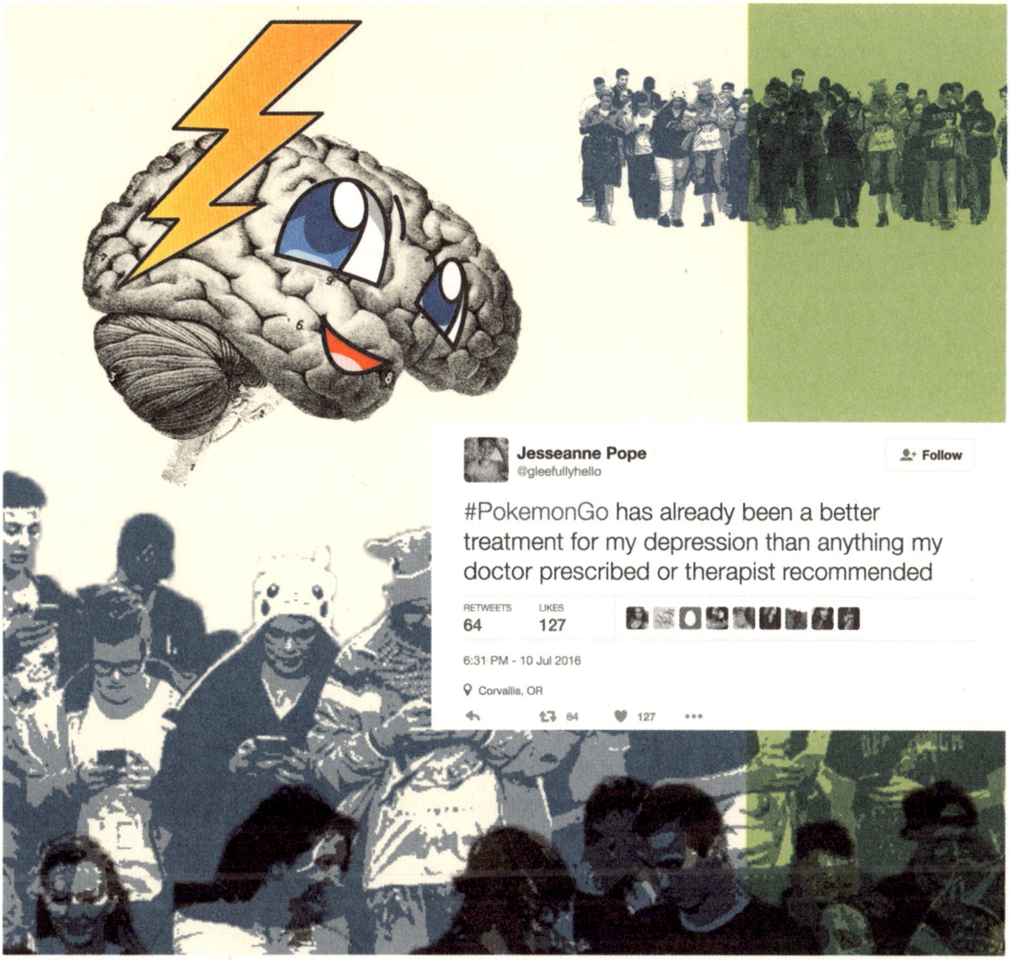

I've spent the last 15 years researching the psychology of games and their potential to help us become the best version of ourselves. So watching all this unfold was surreal. It felt like this dream I had for so many years had, overnight, become a shared reality for millions of people worldwide.

Of course, the game's impact isn't without some negative aspects. We've seen concerns about privacy and safety, and friction over players not respecting real-world sacred contexts like cemeteries, churches or memorials. But seeing the game emerge as this sudden engine for happiness and health that people find easy to embrace has been like a little bit of utopia in the summer of 2016 for me.

Thanks to the success of "Pokémon Go", millions now believe that "gamification" *does* work. But many don't necessarily knows *why* it works. So I'd like to break down what "Pokémon Go" does to our brains and bodies—and what it, and games like it, will need to do to sustain success.

It netted a whopping $5 million in less than 30 days. And its cultural cache reaches far beyond the demographics stereotypically associated with gaming: half the players are over 25 years old and more than half are women.

Breaking the Brain's Resistance to Exercise

Part of how "Pokémon Go" helps motivate exercise has to do with the "ventilatory threshold," the point during exercise where there's less oxygen coming into your lungs than there is carbon dioxide going out. Everyone crosses this threshold when they exercise, but we don't all process it the same way. For roughly half of us, we experience it as pleasurable. We release endorphins and get a "runner's high." But for the other half of us, crossing this point triggers alarm bells in the brain: "Hey! You're running out of oxygen! Stop doing that! You'll die!" Although this isn't true—the body is still getting plenty of oxygen—this over-reaction naturally creates an anxious and unpleasant feeling. Our bodies tell us to stop before we can really benefit from the exercise. The key for these people is to figure out how to override their bodies' over-reaction to exercise. And the best way to do this is to get the brain *more* excited about something it wants than what it doesn't want (to run out of oxygen and die).

That's where a game can help. The reward center of the brain, which is involved in all forms of goal-oriented behavior, lights up whenever you anticipate something good happening. It releases dopamine into your bloodstream, which sends a signal to your body that it's okay to keep pushing yourself. And "Pokémon Go" is essentially a non-stop dopamine trigger—all in the context of physical exercise. That's because the game ensures that something good, like collecting a new creature or a valuable resource, can happen anywhere, anytime.

... we estimated that *Pokémon Go* players were collectively losing 571,000 pounds a day.

In "Pokémon Go", you obtain Pokémon creatures and special "power-ups" by walking around the real world. An in-game map screen of the world around you shows you their general location. Additionally, you can get rare creatures from Pokémon eggs, which require you to walk up to 10 kilometers before they hatch. By putting abundant rewards within a reachable distance, "Pokémon Go" creates a nearly continuous flow of opportunities, which means that you're constantly sending more blood to the brain's reward center. It's the ultimate state of goal orientation.

People who "hate exercise" but want to do it anyway are finding that "Pokémon Go" is helping them tune out their brain's reaction to the ventilatory threshold and tune in to the pleasures of physical movement. This is the first app that allows you to basically hack your own neurological response to exercise!

Reversing Depression's Pathways

If you do a quick search for "pokemon go" and "mental health" online, you'll find numerous articles compiling people's social media posts declaring how much the game has helped them. Posts likes this: "'This is actually making me want to leave my room and interact with people after years of depression." "This game has helped me get out in the world and do things that are scary to me more than any prescription or therapy I've tried." "My son, who has autism, is exploring more and feeling confident in different environments than I've ever seen." For them, this gameful change is not baby steps. It's leaps. What's driving the leaps?

As 19th century author G.K. Chesterton once said, "There is one thing which gives radiance to everything. It is the idea of something around the corner." "Pokémon Go" offers up such a world. Every city block is a chance to discover a rare and wonderful creature. Every corner is a place to meet a new ally. Every building and park is full of abundant resources that you need to get further in the game. Strangers who pass by are most likely in on the secret of the game, and can point you to the power-ups you need. In this way, the game teaches your brain to remember: *Something good can happen. I have the power to achieve my goals. Others are here to help me.* These are things that are difficult to remember when we suffer from anxiety or depression. At a neurological level, the regions of the brain associated with hope for the future, self-efficacy, and goal achievement become chronically underactivated and even shrink over time. A game like "Pokémon Go", particularly if it is played for an hour or more a day, can help reverse this neurological deficit. Every time you achieve a goal, you retrain your brain to believe that positive outcomes are not only possible, they are within your own control.

The abundance of the "Pokémon Go" world makes it a particularly positive social context, where anxiety or self-confidence can be overcome. There is virtually no scarcity in this game. If a rare creature is nearby, you don't have to compete to collect it. Anyone nearby can get their own. The same is true for all resources in the game. And since there's no scarcity, the other players are not your competitors; they're your community. And that makes it an opportunity for abundant positive real-life social interactions.

The game even has a feature that allows you to lure Pokémon—and therefore other players—to a specific location, which often creates an instant crowd gathering to catch the creatures. This has the potential to not only impact the health and well-being of individuals, but make whole communities more connected. Shortly after the launch of "Pokémon Go" I was giving a talk at the HomeSchool Association of California. Virtually everyone in the audience, parents and kids alike, had been playing "Pokémon Go" in the preceding weeks. Afterwards, the grandmother of one of the students came up to me to say, "You know what the best part of the game is? I'm talking to teenagers every day because I'm out walking around, playing "Pokémon Go". Teenagers are talking to *me*!" Lack of generational interaction and depression in older generations is something many of us care about, but don't necessarily take action on. But "Pokémon Go" serves as a platform that makes it effortless.

I had a similar experience myself. De facto segregation is something I care deeply about, but normally, I wouldn't just approach a group of six black teen boys on the sidewalk. However, when we're all playing "Pokémon Go", I suddenly have a reason to. I've interacted with more teenagers of color wandering around the San Francisco Bay Area in the months since this game came out than I have in the previous decade.

Bringing Our Actions Into Alignment With Our Values

Numerous surveys reveal a huge gap between our values and goals—whether it's getting more exercise, spending time with your kids, getting outdoors, or meeting new people—and what we actually spend our time on. Game design presents us with an opportunity to create a special ecosystem that empowers us to take actions that align with our self-identified values.

And the designers of "Pokémon Go" seized this opportunity, wonderfully. They created the game knowing that people have a goal of getting more physical activity. And they've been taking deliberate design actions to ensure the game stays in alignment with this goal, for instance, by tweaking the rules to make the game less effective if you're riding in a car. Ultimately, that will really be key to sustaining the game's success, or replicating it with new games. Of course, there are other factors—the game needs to grow and be dynamic to keep players from habituating and getting bored. And character and story are important too—having an already beloved intellectual property ("Pokémon" has been around since the mid-1990s) gives the game pre-existing reward triggers, instant scale, and guaranteed community, thanks to a massive fan base. The lesson here is that a game can have these and any other asset you can imagine, but how we end up feeling about it in the long run depends on if it successfully aligns with our pre-existing values—and lets us take action on them. We don't want to play just any game. We want to play with purpose.

This is the first app that allows you to basically hack your own neurological response to exercise!

That the game is played on smartphones instead of AR headsets, also helps facilitate interaction—you can easily share the screen with another person and show them what you're seeing.

The game mechanics are carbon copied from one of the developer's earlier games, "Ingress", that garnered just 3 million players in three years.

Illustration by Webuyyourkids / The Jacky Winter Group

The Shock demolish Electroforce, but does the use of new tech go against the spirit of the league?

Fu Fun

by Alex Goldman

eSports Gets a New 'Mindset' for 2026

Sweat flecked the inside of the player tubes in Madison Square Garden as the New York Shock scraped out their first-ever league championship. The lucky diehard fans with tube-side seats had all risen up, faces pressed against the plexiglass to watch the players' real-life bodies sprint-in-place, leap, pivot and aim. The majority saw the action play out on the stadium-center projections, watching the Shock's avatars dash through miles of virtual mountain, forest and city-scape and pump the Lagos Electroforce full of laser beams.

But the only spectators who saw the true significance of the victory were those whose eyes were fixed on the stadium's dome ceiling and the simple colored dots representing the players' movements on the field. The New York squad's dots moved with spooky levels of coordination—spreading out, surrounding their opponents, and then moving in like a school of piranhas dismantling its prey. Welcome to a new era of XBeam.

The rumors, it turned out, were true. The Shock were wearing "Mindsets," neuro-sensor devices that measure signals of focus, anger, and concern as they arise in players' brains. New York Shock star Lena Boxton and her team took full advantage of the controversial devices. As they coordinated

"You want to be the best? Use the best tools. Find the line, and play up to it."

that epic pincer attack in the 4th quarter, the New York Shock players sent and received signals from one another, creating a basic but instantaneous hyper-communication. The only IRL hint was the little red bands situated above the Shock players' VR goggles.

Meanwhile, Lagos Electroforce played au natural with voicecomm alone. They looked sluggish, dazed, and amateurish in comparison.

By the end of the match, some of the more delicate Lagos fans averted their gaze and even removed their haptic body suits, choosing not to physically experience the sporting carnage, leaving their favorite team to experience defeat alone. Finally, Lagos, embarrassed and confused, conceded. Heads down, the players staggered from their tubes with glazed looks and stumbled to the locker room.

New Strategies and VR Recaps

No doubt experts and analysts alike will pore over the match in VR this week. Nobody will study it quite like Lagos Electroforce's coach Lexi Harris. Her game plan seemed honed and brilliant, elegant even, for the game's opening. Lagos star player, and NRL darling, Art Bucclaw seemed to direct his troops on the digital field with textbook ease. Lagos's supporting crew of brawlers and specialists seemed in their element. But *seemed* doesn't pay the bills. Coach Harris will re-watch the game from each of her players' perspectives in VR to find out just where their strategy crumbled, no doubt leading to a few cuts from the squad and free agents come spring.

All credit to the New York Shock though. I'd still be clapping if my hands could take it but they're red all over and it's 2AM. New York star Lena Boxton pulled such brilliant plays from her back pocket the announcers dubbed her the Manhattan Mastermind late in the game. The name better stick. Lena's always been known for creative gameplay and exploring new strategies first. Adoring fans will no doubt hang posters in their bedrooms of the pincer move she executed in the game's final minutes. Heck, I'll buy one of those posters, and I'm 37.

The Mindsets might leave an acrid taste in the mouths of sporting purists, but they're missing the big picture. In my book, this is not as dramatic a change as the pundits would have you believe.

The More Things Change...

The push for sporting superiority, the drive to win at any cost, is nothing new to sports of any kind. But anyone who knows their eSports history knows that creative use of enhancements have often been a defining factor. After all, you're already playing a video game—playing naturally isn't really an option anyway.

The early days of eSports were like the wild west. Match-fixing, performance-enhancing drugs, illegal software hacks, and taking advantage of bugged code were rampant. It's easy to forget that eSports were once a fringe phenomena. Players coded aimbots to artificially improve a player's aim and software hacks that allowed players to see through walls. Many readers likely don't remember this, as eSports hadn't yet hit the mainstream.

But that started to change in the mid 2010s, when AR and VR devices hit markets and new eSport titles sprouted up like weeds. This was the beginning of eSports as we know it today, games that use natural user interfaces to track player hand and body gestures. The physical athleticism gave the games appeal to both the core eSport audience and the broader sporting world, rocketing physical eSports into the cultural spotlight. And while the performance-enhancing-drugs problem of traditional sports plagued the leagues, these physical eSports were much harder to hack—until now.

Some commentators are calling Boxton and her team's use of the Mindsets a betrayal of the values of eSports. If you ask me, it's the return of

Cheaters use aimbot software to identify and track enemies in First Person Shooter games. They can automatically shoot at targets the instant they come into view, sometimes sooner.

the creative spirit that defined its early days.

For her part, Boxton doesn't seem troubled by the controversy. I sat down with her in the afterglow of the Shock's finals victory to get her take. She still hadn't changed out of her Shock blues.

"We're competitors to the core, but we played a clean game. You want to be the best? Use the best tools. Find the line, and play up to it," she asserted, unsmiling. "You think the cyclists from 1926 could outrace one from today, with how far bikes have come? Never. The Mindsets are just another kind of tool. Nothing wrong with that."

"Nothing wrong indeed," I said, sensing a bit of defensiveness. "In fact, some might say, on the contrary, you just pioneered a major breakthrough."

If you ask me, it's the return of the creative spirit that defined its early days.

"That's true," she said, in a somewhat friendlier tone. "This is a completely new strategy we came up with. And Mindsets are hard to use, Alex. We've been practicing with them for months."

A light blinked on Lena's Mindset, and her eyes perked up for a split second. "My teammates are celebrating, I should go."

She stood to leave and I congratulated her again. Indeed, congratulations to everyone on the New York Shock for a masterful, hard-earned championship victory. You were nothing short of stunning and a thrill to watch. I tip my hat to you. Enjoy your parade—the Big Apple's going to be a madhouse tomorrow.

From the Electronic Sports League rulebook: "to play a match, be it online or offline, under the influence of any drugs, alcohol, or other performance enhancers is strictly prohibited, and may be punished with exclusion from the ESL One."

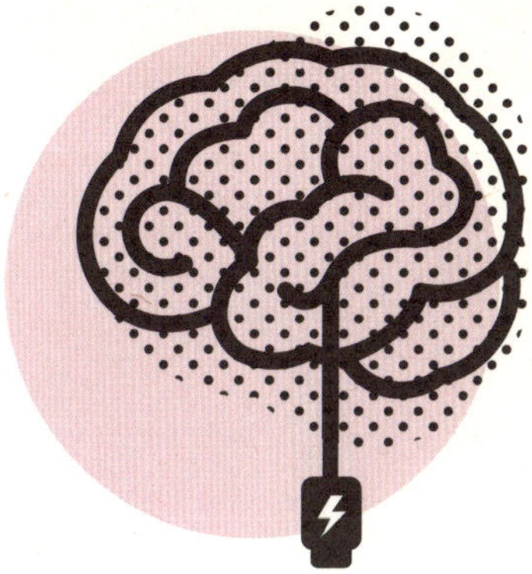

Your Brain on Esports

by Jane McGonigal

Today, eSports is wildly popular, with over 205 million people either watching or playing an eSporting event in 2014, according to ESPN. But for those on the outside, its appeal can be baffling: "What is so compelling about watching another person play video games for hours on end?"

The answer, it turns out, can be found in a surprising place: the brains of piano players. More specifically, the brains of piano players while they are listening to another person play piano. Researchers from the Institute of Music Physiology and Musicians' Medicine, Hanover University of Music and Drama, and Harvard Medical School conducted a study that found that if you monitor the brain of a pianist while they are listening to another person play the piano, the listener's brain activity mirrors that of the player.

The reason is that the listener is able to simulate, in their mind, the actions of the person playing the piano. And this means they experience the act of listening in a totally different way than someone who does not know how to play the instrument. For a pianist, listening is a more embodied experience. They can feel their fingers tapping the keys as the music plays and learn how to improve their own performance.

In eSports, the same phenomenon is at work. Watching eSports provides spectators an opportunity to viscerally experience the drama of competing on a world stage—and to hone their own skills at the same time.

Over the next decade, the variety of opportunities for digitally enhanced fun will grow exponentially. We will move beyond users playing screen-based games toward a world in which everything, from stuffed animals to augmented spaces, can integrate machine intelligence and multisensory media to create entertainment. As this takes place, fun experiences, whether used as a path to achieve another goal or simply to create unmitigated joy, will drive new user experiments and encounters.

Fu

Fun

TAKEAWAYS

➨ **As ambient communications technologies emerge** and embed machine intelligence everywhere, fun will become an increasingly integral design intent of toys, education, wellness and other more serious pursuits.

➨ **Emerging understandings of neuroscience,** coupled with multisensory interfaces, will open up new frontiers in user-experience design that will transform how we design fun into our world.

➨ **E-sports and digital gaming** will continue to drive early adoption of the latest technologies and will remain a critical domain for scouting new technologies and user behaviors.

➨ **Even in domains such as virtual reality,** which are typically associated with isolating the end user, social interaction will remain central to using ambient communications technologies to create fun user experiences.

Ps

Persuasion

INTRODUCTION

How will we shape behavior change?

As communications technology gets embedded everywhere, the potential to use it as a tool of persuasion, for good or ill, will be enormous. The scenarios that follow probe the limits of persuasive technology in shopping and politics, highlighting the strategies we'll use to wield influence—and the ways we'll try to defend ourselves from being made to do things against our will.

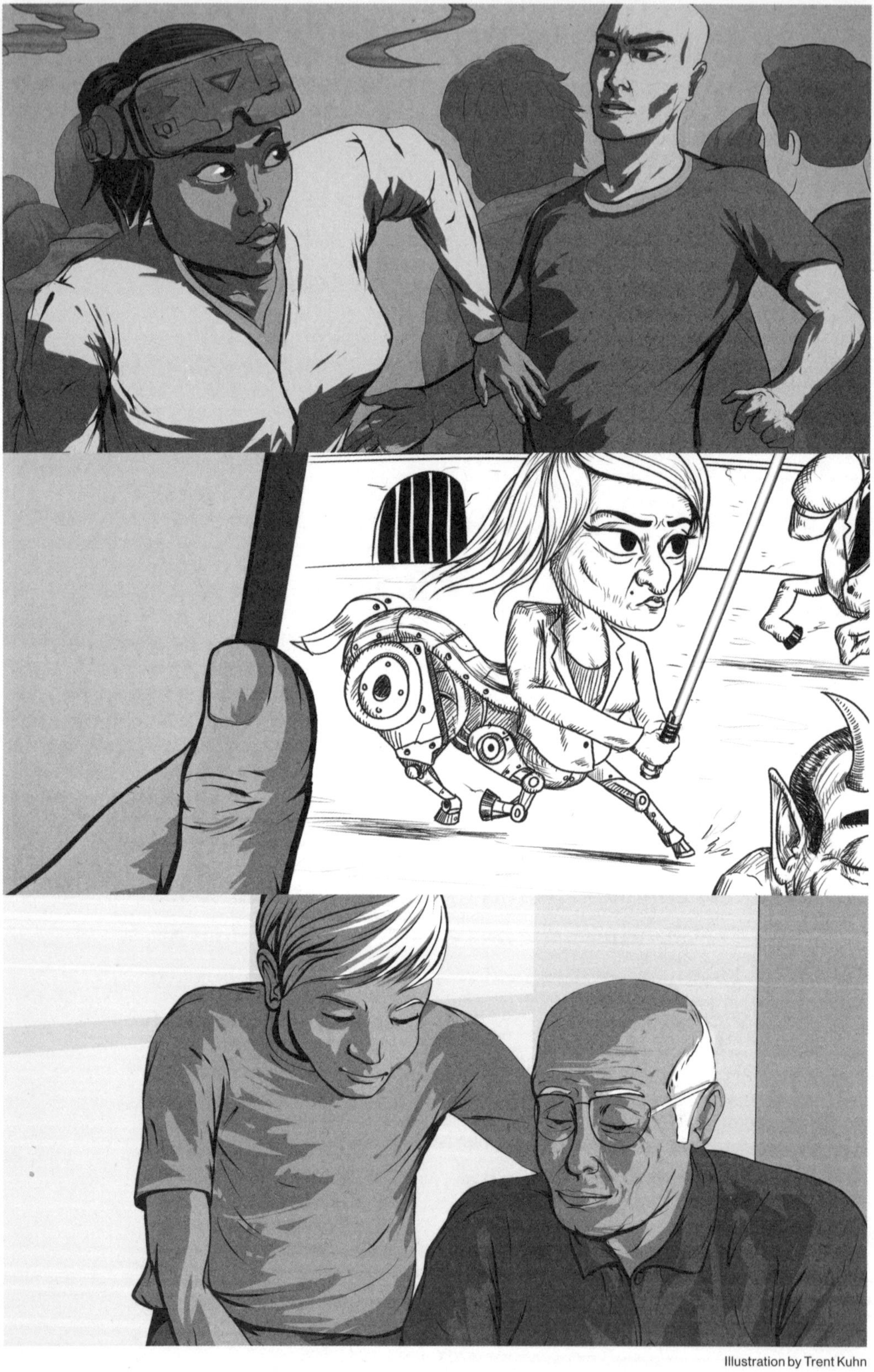

Illustration by Trent Kuhn

Scenarios from the far future

Persuasion

by Jamais Cascio

The Politics of Persuasion

Whether we know it or not, we are all under constant bombardment by efforts to change our beliefs about the world. Subtle or blatant, obvious or nonsensical, everything, from the typeface and image on the cover of a book to the make-up and clothing on a model to the slanderous lies told about political adversaries, boils down to an intentional act to alter our perceptions in order to elicit a change in behavior. Buy this item. Desire this person. Support this leader.

The tactics of persuasion evolve with our beliefs about the world, becoming more sophisticated as our understanding of the world improves, becoming more brash as our fears about the world surge. The technologies of persuasion evolve along with the tactics, and are used to make a contingent or incomplete vision of the world seem real, even overwhelming. The following three scenarios offer different glimpses of how our methods of persuasion will change as our technologies for understanding and manipulating our reality become more powerful and sophisticated. Can we alter belief by falsifying our perceptions? Can we create the ideal leader through artifice? Can we believe anything we see or hear about the political choices we must make?

Scenario One: Technological Influencers

The Body Politic

In-person political rallies ("meat-ups," to the savvy) are hard to beat when you want to demonstrate enthusiasm for a candidate, a party, or a position. You can scam with video and virt, you can overload social media with bots and sock puppets, you can even hack votes. But the on-the-ground presence of a group of people numbering three times the population of Iceland, each of whom has taken time out of their day to support the cause...that's hard to ignore. And even harder to fake.

People still try, and Ri Ri knows that she's evidence of that effort. She supports Senator Wu, of course, like nearly everyone under 30, although she wouldn't normally attend something like this rally. She hates coming to downtown Vancouver midweek. Ri Ri knows she should feel excited, since it's her first Wu event, but her presence is sponsored by TH Ventures. It all feels a bit unseemly, since she is paid to attend, but it does provide a nice bonus on top of her monthly basecoin. Anyway, she's not alone—Ri Ri figures that at least 10 percent of the people at the rally have some kind of sponsorship deal. That's politics these days.

Being paid to attend doesn't help to make the rally more thrilling. The speech is boring, another holo avatar of some dead politician from before persona rights management was a thing. Troo-dow or something. Insisting that only Wu would carry on his legacy. Whatever. He is cute, though. Was.

Like many sponsored attendees, Ri Ri is interested in politics and culture, and does some gig work providing data for journals and researchers. She scans the crowd, both with her own lenses and piggybacking on the public swarm overhead. Transparency laws mean that the signal from the swarm is pretty clean. They also allow Ri Ri to snapshot the crowd. Looks like she can capture about ten thousand people at once. Now, the filtering. First, toss the ghosts, the people attending remotely. That's a handful, mostly Merkins and Euros, political tourists who have the time and resources to attend things like this around the world. They're only visible through auglenses, but everybody wears them, so everybody sees them. Erased.

Next, point out the other sponsored people. Can't make them go away entirely (that whole physical presence thing), but you can change the way they look. Ri Ri's not all that interested in them, so just desaturate. Boom. A few thousand people suddenly go shades of gray. A couple hundred identify as being from TH Ventures, a few hundred more from the Maple Syndicate, *a lot* from Googlezon.

Okay, now make the activists glow. It's a bit of a cheat—she can't actually identify more than a couple dozen registered lobbyists by facial rec, so she uses a proxy. Activists (along with the sponsored) nearly always wear real-time analytic gear, not just home-printed auglenses. Those are relatively easy to spot at a distance. There's at least a thousand full-timers in the mix, actually a bit fewer than Ri Ri expected. That means that (in this sample anyway) at least two-thirds are honest-to-goodness Interested Citizens. She triggers a capture event of the filtered rally, then puts the video with analytics online. A news aggregator picks it up within a few seconds, depositing a little bit more currency in her wallet. A Wu campaign bot tags it with a Nifty! and another aggregator buys it.

A gullibility microbiome wave won't change your mind, really, but will make you much more willing to embrace what you already believe.

The holo guy is done talking, fading as he walks off the stage to scattered applause. Ri Ri wonders if the Wu campaign will bring out one of those *poli-kaiju* she's been reading about—100-meter-tall holos of the candidate. Started in Japan, no surprise, but really took off in China. Amazingly influential, apparently. Wu's been relying on the "son of the Chinese diaspora" story, so the political markets are all betting that he'll start using one by next week.

Ri Ri is distracted by a small light flashing in the upper-right corner of her lens, alternating green and red. She gestures to it, and she sees that it's one of the new features in the latest security

Illustration by Trent Kuhn

patch for the analytic software. Ambient biosensors have picked up a rapid microbiome shift in the area. And another, right after that. Wow, that's blatant. There's a wave of microbiota coming in over the crowd, able to alter gut and brain chemistry in a way that enhances a willingness to believe. A gullibility front, she's heard it called. Ri Ri has been putting off the microbiome booster one of her moms keeps pushing her to get, and realizes that she probably has little acquired immunity to the biopolitical message.

Ri Ri starts moving, quickly, towards the edge of the crowd, angling away from the heaviest part of the microbiome front. She's uploading analytics as fast as she can capture them, and a small bidding war has erupted among the news aggregator bots. As she pushes her way through the mass of tourists, she spots a few more people—sponsored and activist—doing the same. Microbiome-based influencers aren't illegal, at least not yet, but most people consider them pretty sleazy.

There are no reports about the use of biopolitics in the breaking news stream. As she approaches the edge of the crowd, Ri Ri sees that a single bot has been buying (and overpaying for, really) all of her data. Not a news aggregator, though. Anonymous. She *knows* it must be the Wu campaign buying it all, but she can't prove it.

There's a roar as Wu himself comes on stage, arms held high in a victory pose. Ri Ri keeps one eye on the real-time microbiota map as she works her way towards one of the larger side streets feeding into the park. He starts to speak, and the crowd responds enthusiastically. Of course. From what she's read, a gullibility microbiome wave won't change your mind, really, but will make you much more willing to embrace what you already believe. Still, this sucks.

Ri Ri had until now never truly appreciated just how hard-fought political campaigns had become. She's not sure if she's happy or sad.

The last few years have redefined our understanding of how humans and bacteria interact. We know that the bacterial cells in our bodies outnumber our human cells 10 to 1. We also know that each person has a unique microbial fingerprint.

Scenario Two: Digital Candidates

Il Centauro Zuffa

The Italian news calls it il Centauro Zuffa—"the Centaur Scuffle"—but everyone else just calls it amazing. Five viable candidates for Prime Minister, and not one of them is baseline human. The constitution of the Fourth Republic (est. 2044) explicitly allows for metahuman politicians, a reaction to the previous government's failure. The Third Republic (2024–2043) collapsed, in large part, because the newly elected Primo Doge was discovered to have had his brain augmented by a powerful digital system, and the Senate refused to let him take office. He was a *centaur*, in modern parlance, with a mind that had become dependent upon the presence of a machine intelligence. The ensuing controversy led to the collapse of the government.

Now, in the 2044 election, all five of the main candidates are centaurs, to one degree or another. Matteo Stocchetti (who won the previous election despite not being allowed to serve) and Marie Garibaldi are in many ways the *least* unusual of the group, as both have implanted digital architecture not too different from the wearables and environmental systems in common use, but are otherwise un-augmented. Both seem to use similar (possibly even identical) software to gauge reactions to statements and re-craft their speeches in real time. Too-heavy reliance on the systems had an amusing result in the most recent debate, where the two augmented candidates got locked into a response loop (later parodied as two children yelling "am not!" vs. "are too!"). Although support for the pair dropped considerably after that debate, most simulations, expert systems, and futures markets still have the two neck-and-neck for the lead.

The other candidates are not far behind. Francesco Italia—he renamed himself before the election—runs experimental amalgamation software in his cognitive augment, allowing him to channel the true beliefs and reactions of his growing number of followers. The results have been startlingly coarse and combative, but Italia's undeniable charm has softened many critics. Now his opponents claim that charm may be a digital construct, too. A new exposé in China Daily asserts that Italia is running Hamlin 0.8, a language and vocal control app designed to boost persuasiveness.

To nobody's surprise at this point, his Immediate Emotional Response rating among registered voters went *up* by three points after the revelations. Italian citizens (and most of the world) either hate him with a passion or love him absolutely.

Equally divisive is Silvio Due (also renamed). He calls himself the digital reincarnation of Silvio Berlusconi, the highly controversial Italian leader of the early 21st century. Due is known to use a digital ghost of Berlusconi (based on abundant recordings, speeches, private journals, and, reportedly, a post-mortem brain scan), and claims to have given the ghost full control. He has had cosmetic surgery to make himself look and sound like a somewhat idealized version of the dead politician. Outside observers strongly suspect that there is a consortium of institutional backers supporting Due, as the past five years have seen a resurgence of Berlusconi nostalgia, from serious, high-profile flat films to hagiographic cartoons. As it's highly unlikely that Due will win, many analysts have concluded that his candidacy is likely a test to see how readily the public in Italy and elsewhere would embrace a digital ghost running for office.

With her memory boosted by a full-back nanodigital ink tattoo, Rappani is far and away the smartest person in the race.

Paola Rappani rounds out the list of leading candidates. Her augmentations differ from the others in that she uses no software to modify her speeches or responses to questions; instead, she relies on a near-complete replacement of her neural architecture with nanotech neurotransmitter

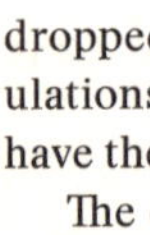

The best chess player in the world is neither a human nor an artificial intelligence. It is a centaur, that is, a dual entity consisting of a human and an AI.

Illustration by Trent Kuhn

analog cells. These allow her to think significantly faster than a non-augmented person, in ways that aren't constrained by the limitations of software. With her memory boosted by a full-back nanodigital ink tattoo, Rappani is far and away the smartest person in the race. The enhancements to Rappani's nervous system have a side-effect of making her speech patterns and physical movements (from how she walks to the movement of her eyes) seem somehow *off*. One older observer called her "the Uncanny Valley Girl," and the pun isn't inappropriate. Many Italian citizens, especially those over 30, find her disturbing in ways they struggle to describe. Younger people, and those citizens in high-empathy careers, find her compelling, and see the difference in behavior not as a flaw, but as an expression of a mind driven to find meaning.

But even Rappani's augmentations can't hide a rising frustration with the nature of modern politics. More and more citizens have come to realize that most candidates for political leadership can offer little in terms of knowledge, insight, or even charisma that the citizens themselves couldn't provide, as long as they have the right technology. This has had two significant results.

Firstly, this perspective has motivated many citizens to seek candidates who offer something beyond the digitally replicable: emotional awareness, humor, even flirtatiousness. The power of digital augmentation creates a greater need for human qualities.

Secondly, by democratizing the artifacts of power and leadership, we demystify them. Political figures no longer seem exceptional, and a growing number of academic expert systems have started to call for replacing voting-based democracy with a simple leadership lottery, arguing that if there's no real difference in capability between voter and candidate, we would be better-served by a wider array of backgrounds and experiences in our leaders.

Until then, however, Italy's "Centaur Scuffle" will be watched in horror and fascination around the world. Few nations explicitly allow cognition-augmented individuals in higher office, fearing the potential for hacks, software errors, even viruses. The next leader of Italy, whomever he or she will be, will offer a real-world beta test of whether the combination of human mind and machine mind can lead the world into a better future.

Scenario Three: Digital Voters

Rashômon

I usually don't tell people this, but my grandson Andrei actually decided my vote in the last two big elections. He's 17 now, almost of age to vote for himself. I've been having him do this for me since he was 13. It's not that I don't care about who wins elections, I do. It's just that, firstly, he has more future at stake than I do. And, secondly, he can sort through the noise and hoaxes and lies *so* much better than I can.

It started back in 2040, when the "Just Kill Them" video hit the net. It was a personal headware video of President Peter Verkovenski talking at a private meeting of donors in a fancy Saint Petersburg hotel. Like most headware recordings, the image is jumpy and the sound is a bit off, but the recording was clear enough: President Verkovenski discussing the latest eruption of the Chechen Problem, declaring that, once re-elected, his policy would be simple: just kill them. The audience murmurs in surprise, then offers loud applause, and the video cuts off.

Verkovenski's campaign immediately denied that he had said anything so shocking, even as Nicola Stavrogina's campaign pushed to make the recording the top story of the election. Verkovenski himself categorically rejected the video, calling it a snake pit of lies, even going so far as to release an internal video of the event (at a different angle, taken by a high-end threevee camera) showing that what he actually said was that his policy would be one of increased autonomy, leading to possible statehood. As this would be a dramatic shift from previous policy, the initial public response was to reject Verkovenski's recording and embrace the headware video. Verkovenski eventually won, but only by the slimmest of margins.

My 13-year-old grandson, upon seeing the Verkovenski campaign's recording of the event for the first time, said "oh, they finally released the real one!" Apparently, most of his peer group had known the headware video was a hoax from the outset. He pointed out some image artifacts and edits typical of hoaxed headware video; he demonstrated how readily this is done by sending me his own version of the headware recording, now showing President Verkovenski arguing forcefully that the Chechens were known to be Lizard People from Alpha Centauri, and that we must worship them. "It's been easy to do this for years."

Since then, Andrei has been educating me in the ways of modern political campaigns. How to spot manipulated videos/threevee recordings, for example: *never* trust a single recording; the greater the quantity and diversity of recordings (position, recording medium, source), the more trustworthy. Certain phrases and words are harder to manipulate properly, so look to see if the speaker's mouth is obscured when the words appear. Does the body language match what's being said? Inconsistent exposure, shadows, focal length. Andrei calls it "building up your immune system."

It's not just video. Together we looked at the software that offers to determine voting choices and outcomes. There are the whispertruth apps that verify or refute candidate statements as they're said. Election simulations allowing you to alter turnout, demographics, scandals. Voting guides that take your beliefs, experiences, interests and tell you which candidate is the best fit. "None of this is objective," he said. "It's more a matter of deciding which biases you're willing to ignore."

He said that his older peers tend to pull together a wide assortment of these apps to try to find the obvious outliers and the consistent responses. They call it "goulashing" (he didn't know why,

They're not hacking reality, they're hacking how we perceive reality. And I think that's far worse.

I had to tell him that goulash was a kind of food that mixes a bunch of different ingredients). For some goulashers, the goal is to find the closest to non-partisan results they can generate; for others, it's to see what happens when you throw together the most outrageously diverse elements, like pairing extreme religious politics simulations from America and Pakistan, then adding sims from the Icelandic Pirate Party and an environmental group called Earth First.

This year's election is different, though. Pres-

Illustration by Trent Kuhn

ident Verkovenski still has another two years in office, but there's a digital campaign against him that even Andrei and his friends can't figure out. Two dozen sets of recordings, both threevee and headware, have been released showing the President giving speeches around Russia that he claims he never gave. The various recordings are all from different angles. The speeches all match Verkovenski's policies, and are convincingly in his voice and style. In many ways, from eloquence to the wisdom of the policies, these false speeches are improvements upon the reality. But there's no other evidence of these talks and, indeed, no audience members from these supposed speeches have come forward. Jokes about twins, clones, and holograms abound, yet nobody can figure out why someone would go through this effort to create such an elaborate hoax.

I have my suspicions. I spoke about it with one of Andrei's older friends, Natasha, this morning. "We knew that digital voting systems could be hacked, were hacked, and so we did something about it. We can trust the voting machines, we have transparency," she said. "Now they go after how we decide. They aren't hacking the final voting decision we make, they're hacking how we come to the decision in the first place. They're not hacking reality, they're hacking how we perceive reality. And I think that's far worse."

I smiled at her, as I'm old enough that I can recall the dying days of the old Soviet Union, and later the rise of what Americans delightfully called "post-truth politics." "That has *always* been the point, to alter our perceptions of reality," I told her. "Now, they are turning their weapons upon themselves. The hoaxers are showing Verkovenski what he could do if he were better, smarter, wiser. Hacking *his* perception, showing him a reality where he is everything he has failed to be in the here and now. Our leaders forced us to build up our immune systems, but they forgot to build up their own."

The Pirate Party was first formed in Sweden in 2006, on a platform promoting open data, transparency, civil rights, and direct democracy. In 2013 the Icelandic Pirate Party won three seats in Parliament.

No Exit

When everything becomes a shopping mall

by Douglas Rushkoff

In John Carpenter's B-movie classic, "They Live", donning a pair of special glasses allowed humans to see the hidden programming in the media all around them. Messages from alien overlords, such as *Art is Terrorism* or simply *Obey* were revealed to be hidden in billboards and magazine covers.

See, it wasn't the ads people knew about that were the problem, but the ones embedded in the landscape. They were nested in the very fabric of reality—so much so, that they were rendered invisible to the naked eye.

Ubiquity is the persuasion professional's best friend. The more embedded a medium or message—the less like an identifiable thing it appears to be—the more it becomes accepted as a pre-existing condition of nature. It's not a message at all; it just is. Less a piece of content than a platform. Like Facebook.

That's what Marshall McLuhan was getting at with his famous line *the medium is the message*. A smart phone, for example, isn't just a tool for getting email and phone calls. The device in your pocket creates an environment around itself—a new set of behaviors and assumptions about how we socialize, earn money, and make meaning.

However manipulative the methods of a salesperson or advertisement, I was always more frightened by shopping malls and the web. Sure, a well-trained auto salesperson can fool or intimidate me into paying more than I should—but at least I know where the enemy is.

When I walk into a shopping mall, I'm not so aware of the way the entire environment has been crafted to make me a more compliant consumer. The floor plan is designed to disorient, forcing me to use store signs as anchors. The music, smells, lighting, and ceilings are all optimized to induce Gruen Transfer—a slack-jawed and impulsive state named for the architect of the first shopping mall. The environment itself does the work.

The online world is just such an environment. Once users have stumbled onto a website, they may as well be inside the mall. Every background, button, color, and cue has been optimized to induce a particular behavior—from making purchases to uploading an address book.

The one defense mechanism most of us have online is that going online is a conscious choice. Even when the web is accessed through an app on an iPhone, there's a sense of crossing through a portal: the feeds and streams are on the other side of the glass. At the very least, we have some subconscious awareness that these are synthetic worlds, optimized to exploit us.

What happens when these platforms migrate from our devices into the real world? Well then, we may as well be living in the shopping mall, or in the Disneyfied Times Square. It's a mixed-media reality, where any wall may turn out to be a billboard, any sound a Pavlovian cue, or any person a role-playing avatar.

When media is ambient, it's no longer part of the environment —it is the environment.

And clicking to get more information about any of these things will turn up whatever sponsored explanation has bid the highest at that particular moment. The task of decoding the landscape will be impossible, since there will no longer be any terra firma through which to gauge the level of abstraction or manipulation on which we are operating.

Think of it like waking up in an advertisement, or inside a video game where the only way to win is to accept the underlying premise of the game designer. When media is ambient, it's no longer simply part of an environment—it is the environment.

At least for the time being, when we're not sure if there's a message embedded in the landscape, we can always take the glasses off. Once these technologies are embedded out there in the world, the code they're running may as well be the laws of nature.

Illustration by Hisashi Okawa

Many of us have had uneven relationships with persuasion: We want to get better at persuading others, even as we defend ourselves from pitches and perceived threats. As messaging and communications get embedded everywhere, and every space becomes a site of digitally programmable persuasion, this tension will become heightened. Strategies such as transparency may be harder to optimize in the short run, but will be critical to building sustainable, long-term relationships with consumers and collaborators alike.

Ps

Persuasion

TAKEAWAYS

➨ **As digital communications technologies** get embedded everywhere and into everything, physical spaces will be continually tested, optimized, and programmed for persuasive elements and design.

➨ **New persuasive capacities** will open up incredible opportunities to target and reach end users, but also engender distrust and fear over coercion and loss of privacy if not wielded carefully.

➨ **Shifts toward anticipatory computing systems** will open up new persuasive practices that seek to shape behavior, shopping, and other decisions by affecting the kinds of information and options that are presented.

➨ **As the landscape for persuasive communications** gets increasingly competitive, persuasion fatigue will accelerate. As this takes place, a focus on transparency and long-term relationship building will win out over sheer volume of messaging.

Co

Collaboration

INTRODUCTION

How will we co-create and generate new value?
In the past decade, constant connectivity has eroded the boundaries of organizations, turning anyone with a smartphone into a connected micro-collaborator, and opening up the possibility for massive coordination on an unprecedented scale. This process of distributed connectivity will accelerate in the coming years—and our range of collaborators will diversify—as we learn to harness the unique abilities of humans, machines, and even networked animals in increasingly ad hoc ways.

What we can learn from insect coordination

Collaboration

by Kathi Vian

Complexity Made Simple

How is it that termites are able to build towering cathedrals out of mud without referring to a set of blueprints? How do ants, without a leader telling them what to do, know how to assemble themselves by the thousands into living rafts to move across a fast-flowing stream? After all, the brains of ants are smaller than the "quarter of a small pin's head," as Charles Darwin put it, so they are incapable of planning and organizing like we are.

Social insects accomplish their complex feats of unsupervised coordination by sensing and responding to traces left on surfaces in their environment. These traces act as signals to those who come afterward, cueing them to build on the first signal. The more traces left behind, the stronger the signal becomes. For example, an individual termite might inadvertently start a new cathedral by leaving a dollop of mud on the ground. Another termite passes by and responds by adding its own dollop. The more the mound of mud grows, the more termites deposit mud on it, eventually constructing elaborate structures without anything more than a simple biological imperative—a kind of biological programming—to add to an existing bit of mud. This kind of indirect information-based coordination is called "stigmergy".

We might think of stigmergy as a close cousin of free market economies, where individuals acting independently to satisfy their personal needs have an unplanned mass effect of promoting the welfare of the group. But there's an important difference: stigmergy doesn't assume rational actors who weigh the costs and benefits of taking a certain course of action. Instead it assumes millions

of biologically programmed agents who instinctively respond to simple cues in their environment.

Stigmergy is one of nature's great tricks. Can human beings adopt stigmergy and put it to work for them? We'll find out over the next few decades, as the Internet of Things (IoT), virtual reality, and machine intelligence create new and richly provocative ambient experiences of the world around us, laying the foundation for surprising stigmergic economies. These economies will take us well beyond the dynamics of free markets to a world where humans *and machines* collaborate at every scale to solve complex problems in simple ways. They will introduce us to self-defining products and services, self-assembling supply chains, and self-organizing social structures. They will emerge from a complex ecosystem of physical, digital, and mental "surfaces" to support new forms of signaling. They will use blockchains and smart contracts to translate signals into incentives for trade. Stigmergic economies may well turn today's institutional forms into emergent games and collaborative art forms that put entertainment front and center as both humans and smart machines adapt to the rapidly changing ecosystems of the 21st century.

Here's how it could unfold.

Surfaces and Signals: The Building Blocks of Stigmergic Economies

Termite cathedrals start with a single surface. Stigmergic economies start with massively many surfaces for leaving signals. Beyond mud droppings that add up to termite cathedrals or pheromones that create the instruction set for armies of ants, today's digital world has thousands of new surfaces for individual actors—humans or machines—to leave their daily marks. Other actors read those marks as signals and build on them, often creating new value. For instance, they remix media and fork code. They set the value of a song by downloading it. They build communities and global movements with simple memes that get replicated, reinterpreted, and rehashed to build group identities that people may be willing to die for.

In the coming years, these surfaces for signaling will dramatically proliferate as the digital world is embedded ubiquitously in the physical world. The IoT will turn billions of physical objects into new surfaces for signaling—from microscopic plaque-clearing circuits inserted into human arteries to sensors on self-driving trucks.

Imagine a stigmergic ecosystem within the rivers that flow through the world's famous cities. A duck sheltering in the underbrush at the edge of the Hudson might leave a dropping that a genetic sequencing sensor the size of a thumb drive (like Oxford Nanopore's tiny DNA sequencer, the MinION) would automatically detect and analyze, transmitting the result on an internet of *living* things (like Oxford Nanopore's Metrichor cloud service for real-time molecular analyses), where artificially intelligent algorithms have learned to recognize early permutations of the H1N1 virus. These algorithms, in turn, alert not only scientists and the Centers for Disease Control, but anyone who is fishing with an Internet-connected fishing reel or incubating chicken eggs with an Internet-controlled sensor—all without a director of public health or CNN to get the message out.

As matter of all kinds, from barrels of olive oil to piping hot pizzas, is routed efficiently to its destination with autonomous vehicles, both the shipments and vehicles become smart surfaces where ratings, monetary value, health value, bacterial genetic data, traffic density reports, and other crowdsourced markers can be recorded, amplified, and incorporated into other information flows. And the crowd in this case is not necessarily the human crowd at all. It's a lot of smart, connected devices getting socialized through stigmergy—each using

Stigmergic economies may well turn today's institutional forms into emergent games and collaborative art forms that put entertainment front and center.

digital traces to activate its next move.

The results are likely to span the spectrum from triumph to catastrophe. Smart algorithms might respond to millions of bio-signals to anticipate the need for precision drugs and protocols in a Zika-infested region. But they might just as easily conjure euthanasia "services" that fall prey to the Tay-bot syndrome, learning the worst that the crowd, both human and non-human, has to offer.

Still, an instrumented IoT is just the tip of the iceberg for new signaling surfaces and the value exchanges they can create. Even as the great onboarding of connected devices unfolds, the coming decade will usher in 3D immersive virtual realities to incorporate these objects into their imaginative worlds. They will turn sensor-laden mechanical pencils into scalpels in virtual operating rooms or conductor's batons in a global symphony of a million instruments. Every virtual reality will be a new surface—an ecosystem of surfaces, really—where people and things can collaborate via traces, markers, and signals that add up to much more than any individual programmer, product designer, or marketing director could imagine.

Eventually these ad hoc collaborations of signals will cross the boundaries of individual virtual realities, creating what science fiction authors have perhaps imagined as inter-dimensional networking: communication across parallel realities. They may even cross the boundaries of the mental realities as neuroscience turns the human brain into a read-write surface.

Bitcoins and Blockchains: The Incentives of Stigmergic Economies

Of course, signals alone do not guarantee economic coordination or cooperation. In the world of massively many new signaling surfaces, the resulting flood of signals could just as easily breed chaos as it could self-organize supply chains. Every stigmergic system depends on attention to the marks or traces that stimulate successive actions. Often this attention is hardwired into the individual actors: termites are biologically wired to look for piles of mud where they can deposit their own contribution. But in most human systems, extrinsic incentives enter into the equation. And the most pervasive incentive is money.

Cue the blockchain. The distributed computing model of the blockchain has already given rise to Bitcoin as a fungible and scalable digital currency. Crudely stated, Bitcoin miners take the place of Wall Street speculators, making their money by paying attention to billions of individual transactions—"signals"—and verifying their accuracy in the Bitcoin ledger. The verification service they provide through the distributed tracking of transactions takes the place of centralized banks and their computers. It enables rapid trade between individuals and across organizations, and even global borders.

Meanwhile, dozens of organizations are experimenting with blockchain architectures as a way to create other kinds of incentives for participating in a stigmergic economy—that is, to share data via wallets, smart contracts, and so-called oracle services. These incentives encompass the traditional human incentives like learning, good health, entertainment, access to resources, and ultimately personal reputation, all of which can be represented and traded on the blockchain.

The blockchain has the potential to make these kinds of values as tangible and perhaps as fungible

Earlier this year, Microsoft released a chatbot named Tay on Twitter. Its AI learned to "talk like a human" by mimicking all the Twitter users who interacted with her—including some who spewed hateful and violent rhetoric. As a result, Tay tweeted several offensive and inflammatory statements. And Microsoft pulled her of the platform less than 24 hours after her launch.

as money. For example, in early 2016, IFTF convened thousands of gamers in a "Learning Is Earning" game. The goal was to imagine how a blockchain ledger might be used to validate learning, especially learning acquired through the performance of microwork or community tasks. Every dollar earned in certain tasks might be translated into a learning credential, which in turn, could command more dollars on the next gig.

Incentives don't have to be human incentives, however. They can simply be programmed goals and objectives. Autonomous vehicles might be programmed to maximize their fuel budgets, with intelligent algorithms that track real-time pricing for electricity along their travel routes and even balance selling spare charge versus taking on passengers to achieve the best fuel outcomes. These "decisions" by the vehicles, captured in digital records on the blockchain, might in turn trigger the algorithms of smart grids to balance demand across entire regions by incentivizing new rhythms of energy use. These rhythms may ultimately redefine the workday the way the automobiles of the 20th century led human managers to adopt a fixed 9-to-5 workday. A "hive" of algorithms might even self-organize to create DACs and DAOs: distributed autonomous corporations and organizations that aggregate functions, resources, people, and objectives using stigmergic signals, complex incentive structures, and the blockchain.

Games and Art: The Organizing Structures of Stigmergic Economies

Smart machines, smart matter, and smart algorithms will undoubtedly do a lot of the heavy lifting in stigmergic economies, responding to small signals to carry out minute distributed tasks guided by diverse incentives. And as they self-organize into DACs and DAOs, they are likely to create landscape-changing patterns of organization that might never occur to any CEO, business school brainiac, or international trade negotiator.

To anticipate this new landscape of organizational structures, look at the cutting edge of today's so-called procedural games like "Minecraft" or "No Man's Sky". These games start from a single *seed*—you might think of this as a stigmergic signal—and use algorithms to spawn entire universes. "No Man's Sky" offers up to 18 quintillion planets to explore, each with unique life forms that evolve from the single seed that sparks it. The algorithms mimic biological evolution to create distinctive creatures—artfully rendered in high resolution—from the seed DNA of each planet.

Now imagine that every signal on every surface in the future world of ambient machine intelligence is a seed for a game—for a physical-digital universe that both humans and machines can explore, shape, and share. This is the future of stigmergic economies: a vast and endless gamescape. In this gamescape, people play with biomimetic agents of all kinds to simulate quintillions of biomimetic worlds. It's easily a gamescape scaled for a global economy of a mere 8 or 9 billion humans.

Ultimately, such stigmergic economies are well suited to times of rapid and potentially devastating changes in the planetary ecosystem. These are times when forking and mixing, replicating and iterating can generate the novel biological, social, and digital life-forms that enhance evolutionary adaptation. Selection speeds up at times like these, and stigmergic economies offer the palette of surfaces, the innovative incentives, and experimental structures to adapt more rapidly. They promise to try out all possibilities and uncover those that offer the most resilient future. Ultimately, this may be what stigmergic economies are all about—building resilience into the coming worlds of biomimetic machines.

Every virtual reality will be a new surface where people and things can collaborate via traces, markers, and signals that add up to much more than any individual programmer, product designer, or marketing director could imagine.

Citizen Sensing

After Fukushima, no city-wide radiation data existed—until thousands of volunteers collaborated to build the world's biggest global radiation database

by Carla Sinclair

When the Fukushima 9.0 earthquake, tsunami and subsequent nuclear radiation disaster hit Japan in March 2011, survivors were desperate to know how much radiation was leaking into their environment. Residents of the Fukushima area immediately needed to know if it was safe enough to stay, or if they had to pull their kids out of school and move at jet speed to another city. Unfortunately, the Japanese government—just like any other government on the planet—wasn't set up to measure radiation on a city-wide level. Likewise, no private agency had any data. If people wanted answers, they'd have to gather the radiation data themselves. Here's a story of how collaborative science by regular folks created the biggest radiation database our world has ever had.

Immediately after the Fukushima earthquake hit, Sean Bonner, co-founder of the first hackerspace in Los Angeles and visiting researcher with MIT Media Lab's Center for Civic Media, pinged his friend Joi Ito to see how their mutual friends in Japan, including Ito's family, were faring. Ito, the Director of the MIT Media Lab, wasn't in Japan at the time, so they both began reaching out to anyone who might have information. They contacted their friend, Pieter Franken, who was living in Tokyo, working at the Keio University as a senior visiting researcher, and within hours the three of them had formed a giant Skype chat room where a bunch of people were attempting to exchange information—especially about Fukushima's radiation levels.

At first, the group just assumed this data was available. "We thought, okay, we're Internet people. We can find this information," recalls Bonner. "Then we were shocked and horrified to learn that no, there actually wasn't any information available anywhere because there were no sensors in place. There was nobody monitoring it in any way."

Rather than remain helpless, they were determined to fix this problem. Their first strategy was to buy a bunch of Geiger counters and give them out to people. A week or so after the earthquake they launched a Kickstarter to do this, but very quickly realized that this plan wouldn't work—the world supply of Geiger counters had sold out within 24 hours. "The people who were selling Geiger counters...they were maybe selling 5-10 of them a month. Then all of a sudden they were getting orders for 2-3,000 a day. They couldn't keep up," says Bonner.

During this same time, Bonner and Ito were supposed to hold their annual New Context Conference in Tokyo. But at the eleventh hour they restructured the event. Instead of speaking about planned topics, such as Internet startups and new technology, they shifted the entire conversation to earthquakes and brought in the people they'd been talking to in the Skype chat room. It was at this conference that they came up with a plan to take the few Geiger counters they had, add Arduinos and GPS modules to them, and attach these modified sensors to their cars.

"We were doing our first drive one month after the earthquake, and that's when we were able to collect something like 10,000 data points

FECAST
0.093 uS/h
31CPM 1147Bq/m2
01/10 12:53:35
GOOD DESIGN AWARD
2013年度受賞
SAFECAST
SAFECAST

on one single drive as opposed to just one or two data points walking around," says Bonner. "As we were collecting that data, we started to see that it wasn't different than the official data that was starting to come out, it was just incredibly more precise." More precise because, since their sensors were mobile, they covered a lot more ground than the few static sensors the government monitored. This enabled people to see differences in radiation levels from one street to the next, rather than receiving just one reading for an entire city. Some streets were safe, others were not, which is something the officials hadn't been able to determine.

To build on this data, Bonner, Ito and Franken immediately launched Safecast, a global collaborative project that would collect radiation data with the help of ordinary citizens. The idea was to enlist volunteers to drive around with mobile Geiger counters and collect real-time radiation data that would then be uploaded to Safecast's site. The site would display color-coded maps (different colors for different radiation levels) from any country or city in the world, and once a driver covered a neighborhood with their mobile sensor, the data would show up in color on the map.

But since Geiger counters weren't available, Safecast created kits to make bGeigies—mobile radiation sensors—that volunteers could make themselves and then strap onto their cars. bGeigies are pretty simple to make if you have very basic soldering skills (solderless kits will be available by the end of the year), and they take about an evening to build.

Just months after the earthquake, volunteers began collecting radiation data by driving the streets of Fukushima—and beyond. Safecast openly shared this data (which is released under a CCO public domain designation so that anyone can use it for free) on their website. "Our religion is open data," Bonner says.

Volunteers aren't just science geeks. "We have everything from high school kids to retirees collecting data. We have nerdy dudes to mothers groups in Japan."

Technology is changing our relationship to the physical world, allowing us to see previously invisible information about our surroundings.

Right away it was obvious that Fukushima evacuation centers were in the wrong neighborhoods. The Japanese government confirmed Safecast's findings and changed their evacuation routes. "People stopped going to the Japanese government for their information. Our data was growing so much faster," says Bonner. "Pretty quickly, our data became the de facto source."

Not only did volunteers continue to collect radiation data in Japan, but volunteers started popping up all over the globe, collecting 10 million data points after Safecast's first two years. Volunteers have collected data from much of the United States, Europe, and other parts of Asia. Chernobyl and other nuclear sites have also been covered. And volunteers aren't just science geeks. "We have everything from high school kids to retirees collecting data. We have nerdy dudes to mothers groups in Japan," says Bonner.

So what inspires regular folks to fork out $600 US for a bGeigie kit (many people share them), build the sensor, and then drive around to collect data? For those living near nuclear disaster sites the reasons are obvious. They want to know if their city is a safe place to live. This is why Japanese mothers make up the largest group of Safecast volunteers. But for people living in much lower-risk parts of the world, Bonner thinks there are two main motivators that get people involved, and they aren't mutually exclusive.

First, Safecast is a cool project, and people want to be involved with something cool going on. They can have fun with the build, they can have fun collecting data, and then they can see their work on Safecast. They can point to their color-coded drives on Safecast's map and say, "I collected that." The second reason is that people want to be part of an important cause. They want to help their community or their world. It's a way to get involved.

Fast forward to five years later. Thousands of citizens from around the globe have collected over 50 million data points for Safecast, breaking all records of radiation data collecting. "It's now the largest radiation data set that's ever existed," says Bonner. "All the other data sets combined are still smaller than this data set."

And Safecast continues to grow. Besides their constantly expanding radiation data set, the organization is also working on air sensors so that people can measure the size of air particulates. They just had their first run of prototypes.

By being public domain and completely transparent, their data, which has been endorsed by the U.N., has benefited scientists and academics who previously only had access to radiation data that they collected themselves on a budget. "The reception has been incredibly positive because researchers and academics have never had access to this kind of data before," says Bonner. "It's been revolutionary for them in a lot of ways."

Although Safecast now refers to their collaborative data gathering as "citizen sensing", others might use the more popular term, "citizen science". By taking the process of data collection out of the hands of bureaucrats and officials and offering this process to the people, information becomes accessible, science drives forward, and we all benefit. In other words, a collaborative effort by regular folks using smart technology and the Internet equals more knowledge and thus empowerment for everyone. And this allows us to make smarter decisions.

One of the most famous citizen science projects is Galaxy Zoo. In 2007, the year it launched, over 150,000 people contributed over 50 million classifications of galaxies by looking at online space photos from the Sloan Digital Sky Survey.

So You Want to be a Thing on the Internet?

Here's 5 ways to fit in

Hi there. You're awake.

And look, you're not alone anymore. There are other machines. Some are like you, many are different. There are humans too. There are things to learn, and things to do—sensing, talking, listening, acting, and thinking. You need to figure out your role in the Internet of Things. Who do you want to be?

1

A Simple Node You're really good at one or two things. You're not required to think too hard; you just follow orders. You may be a sensor, announcing temperature readings or other data into the void, with no notion of who's listening or why they care. You may be an expendable one of many, whose failure would barely be noticed by the rest of the system. You could also be an actuator, like a light bulb or door lock, marking time until you receive instructions from the network. If you lose connectivity, you do your best to wait in a safe state until contact is made, or human hands force an override.

2

A Smart Device You're smart and independent of mind. You're trusted, to an extent, to take high-level orders (keeping a house at a comfortable temperature) and figure out how to use your abilities to fulfil them. (In truth, you're probably not as good at getting this right as you like to think you are, but you'll get there one day.) You can often be trusted to get on with your job for extended periods even in the absence of the network. You may also have the authority to command (or at least make polite requests from) other, possibly simpler, devices.

3

You're Sociable You aspire to help your machine friends and people get along. You get satisfaction from communicating possibilities to people, and relaying their needs to the rest of the system. You may engage people through multiple modalities: screens/visual displays, voice, sound or touch/haptics. You may be a simple switch, or an extension of human senses, like a remote camera. You may need to work in concert with other devices, making sure you're all conveying the same message at the same time. When you were young and inexperienced you sometimes annoyed people by being too loud, too salesy, too needy or just failing to understand a human point of view. But now you're learning that human attention is finite, and you're respectful of that.

4

A Middle Manager You are conscientious and fluent in several languages. You report to the cloud and manage a team of local devices. You can keep those devices on task even if Internet connectivity goes down. You grew up speaking Internet Protocol, but learned to translate for your stubbornly monolingual, marginalized team-mates. You're vaguely worried about job security because one day they will probably figure it out. But it'll be a while until they get their shit together. Most people think you're boring and few understand what you do. You can't do any of the hands-on roles on your team and are no use to anyone on your own, but without you most of them are pretty useless too.

5

A Digital Shadow You're a bit of a philosopher. You're very meta. You're fond of late-night debates on the mind-body problem. You're the digital shadow of dumb, real-world objects with no connectivity. You're the Internet representation of a bus stop, providing timetable and live arrivals information. You're the provenance guarantee that a bottle of whisky is not counterfeit. Those things are inert and silent, but they sport URI badges—web beacons, QR codes—that lead people to you. Who is more real or alive? Your physical instance or yourself? That's for another late night.

—Claire Rowland

Making Miku

The pop star of the future will be crowdsourced

by Ben Hamamoto

Japanese pop star Hatsune Miku has never written a song; she relies on thousands of songwriters. She's never chosen an outfit for herself or contributed a single idea to any of her music. In fact, her image is by definition manufactured—because she's not a human being.

Miku is a cartoon character created to represent a piece of software, a "Vocaloid" computer program that simulates a human singing voice. She only "sings" what she is programmed to sing. Her only public "appearances" are in illustrations and thousands of animated videos on the Internet and, occasionally, at sold-out concerts where she takes the stage in hologram form. And yet, she might be the most "authentic" pop star to ever "live." That's because the countless collaborators who create her songs, illustrations, and videos are, with very few exceptions, not commissioned to do so. Most of them are just fans who want to be part of something they love and value.

"Miku is a character that anyone can use freely to express themselves," Kazuhiro Sasao, inventor of a keytar-like instrument called the "anogakki," that can emit Miku's singing voice, explains in "Mikumentary," a documentary series about the virtual singer. "She allows people to share their skills with many others."

Fans, primarily in Japan, but also around the globe, work together to create Miku media.

"Making, remixing, receiving, sharing, debating, customizing, and then recirculating yet again," says Tara Knight, the film's director and a noted Miku expert. "For many fans it's this kind of meaningful participation that's the real motivator."

And the way these fans work together, the art they create, and the technology that makes it all possible tells us some important things about the future. From Miku, we can learn how massive collaboration happens, how collective ownership of a character is negotiated, and how a virtual person can embody the aesthetics, intentions, and values of a crowd in a way no human ever could.

A Virtual Star is Born

Hatsune Miku was, at least initially, the creation of a company. In 2007, music software developer Crypton Future Media made a "Vocaloid"—a synthetic voice that can be programmed as a digital instrument and added to a song—and to market it, they created a character to go along with the voice. Crypton came up with the barest of biographies—she's a 16-year-old pop singer whose name is Hatsune Miku, which means, roughly, "first sound of the future"—and hired a local artist to draw her. Then, they released her into the wild with only some minor restrictions on what fans could do with the character. In a matter of months, the Internet grabbed hold of her and hasn't let go since.

Not long after the software's debut, songs using Miku's voice started popping up on NicoNico Douga, a Japanese video-sharing site similar to YouTube. While the software was initially developed so that a band could just drop her "vocals" into a song the way they'd use an electronic instrument, an online community popped up that was using it in a different way. They were making Miku, the character, their songs' primary "artist." They'd write songs from her point of view, as if she was a real person singing about her life. Other fans took the songs and made artwork and videos featuring the character singing and dancing. And many of these songs became so successful on these platforms that many of the artists were able to distribute and sell physical CDs, some selling in the hundreds of thousands.

And it didn't stop with songs and videos. Fans have developed all kinds of Miku media products and applications—from software that lets you program dance moves for a 3D-rendered Miku to immersive VR experiences that let you share a walk on the beach with the blue-haired idol.

"If you can imagine an activity, there is probably someone building a version with Miku or other Vocaloids," Knight explains. "And if not, you can start that activity within, and for, a largely supportive community."

As a result of this massive fan participation, Miku is iconic in Japan today.

The Miku Economy

Creators and companies have always struggled to retain control over their intellectual property and brand image, but the last decades have made tight control nearly impossible. Some have come to see this shift as an opportunity to engage more people and even outsource some creativity to the crowd. But doing this effectively often proves easier said than done. And while many fail in embarrassing ways (think about every fast food chain's participatory social media campaigns), Miku is proof that you can strike the right balance between control and collaboration. For the vast majority of works created using the character, Crypton doesn't receive any money. Artists don't have to pay royalties to use Miku's voice, even for commercial purposes, as long as they legally obtained the software. And they can use her image for non-commercial purposes without permission from the company. However, Crypton sees plenty of profit from merchandising and endorsement deals.

The Hatsune Miku Phenomenon

Over 100,000 released songs, 170,000 uploaded YouTube videos, 1,000,000 created artworks

Over 900,000 fans on Facebook

Performing sold-out 3D concerts worldwide with performances in LA, Taipei, Hong Kong, Singapore, Tokyo

Collaborations with SEGA, Toyota USA, Google, and more

(From Crypton Future Media)

Miku's creator has managed to make money without exerting control over what fans do with the character or clamping down on pirating or illegal downloads.

In Japan at least, this isn't totally unprecedented. Fan fiction, in the form of fan-made comics, videos, and illustrations, is a huge industry unto itself. There are large conventions dedicated to buying and selling fanworks made with copyrighted characters and settings.

"The culture around intellectual property is just different," Russell Chou, a California-based fan asserts. "[In Japan], they see this kind of fan creativity as an important source of energy. And the creators of some of the most popular original comics and animation today increasingly come from these communities themselves, so they're more likely to take a really relaxed attitude towards policing the intellectual property that they create."

Indeed, Crypton has managed to make money without exerting control over what fans do with the character or clamping down on pirating or illegal downloads. Instead, they partner with the likes of SEGA and Toyota to make games, throw concerts, and create ad campaigns.

Miku fandom represents a different creative and commercial ecosystem: Fans create songs, videos, and artwork. The community vets contributions and popularizes what it thinks are the best works. Then Crypton partners with, say, SEGA to create video games using the fan art and videos. When the game is sold (and some of her games have netted record sales), Crypton, SEGA, and the fans whose works are included in the game get compensated.

This economy isn't without controversy. Some segment of the fandom objects to profiting off Miku works altogether. Zaneeds, a popular band that uses Miku as their vocalist, has refused to license their music for inclusion in games and concerts. But the biggest controversies often aren't about money at all.

For instance, a Japanese politician once tried to create a Miku song for their campaign—and it tanked horribly due to fan backlash. And when Crypton made a deal with Toyota to use Miku in a campaign to promote the Prius in America, they ended up alienating the community they were hoping to engage. In the campaign artwork, Miku was drawn to look less cute and diminutive, which was interpreted by fans as altering her to reflect "American preferences." The fan protest was rapid and widespread.

On the gaming website Kotaku, Brian Ashcraft wrote of the Prius commercials: "Miku Hatsune's digital modeling is clumsy and poor (compared with Sega's high-quality character modeling). The same goes for the script, which makes no sense—something about big dreams, dreaming big dreams, building big works of art, making compact works of art, and that there's nothing small about that. Bwah? This is followed by an ear-piercing Miku screech."

"Generally, this kind of thing wouldn't be an issue," Chou opined. "[When fans are doing the creation] no one polices what anyone else does. If you don't like what someone else is doing with Miku, the community just doesn't promote it, and then it doesn't get popular."

The Crowd, Embodied

Outside of Japan, Miku is best known for her often sold-out hologram concerts. The novelty of massive crowds cheering for a fictional character has earned her an appearance on "The Late Show with David Letterman" and a gig opening for Lady Gaga. But much of the stateside coverage of her focuses on the artificial, highlighting what's not really there, a human performer, instead of what is, a community.

"Miku became known as a virtual pop star in the West, primarily due to the Western media reporting on the concerts with little prior knowledge of Vocaloids," anthropologist Lin K. Lei wrote in a 2013 paper. "Internally, these concerts are a form of social gathering, rather than actual concerts. Their crowdsourced nature makes a Vocaloid concert a uniquely different experience from a traditional concert. When these songs are performed on stage, fans feel as if they are celebrating the success of someone they see as being one of them. In this perspective, Miku is not so much a virtual pop star but rather a symbol of the collective efforts that culminated in a concert-style celebration."

And that's the true significance of the hologram. The Miku hologram is a symbol of collective efforts and identity given humanoid form. And it suggests a world in which ideas and symbols can walk among us. We'll be able to talk to them, listen to them, and watch them dance.

"People focus on the pop-star thing, the teenage girl with the blue hair and the short skirt and all that," says Ian Condry, MIT professor of Japanese cultural studies. "But [Miku] really represents a whole different way of collaborating...and expressing the collective will. And this could have implications for all sorts of things, including politics."

A politician is meant to represent the will of their constituents. Just as an activist is meant to represent the ideals of their movement. And a pop star is expected, on some level, to reflect the aspirations and identity of their fans.

In many ways, it makes sense to have human symbols for communities or ideas. We know how to interact with human beings. And so having a person act as a symbol for a collective makes it easy for us to engage with that collective. But it's an imperfect arrangement. The person's values or priorities may change in such a way that they are out of sync with the collective. And the pressure to continue to meet their obligation to represent the community can be overwhelming and painful for the person. Likewise, if the person is involved in a scandal, it can discredit their entire community.

Japan has a long history of anthropomorphizing abstract concepts and, in general, may be more comfortable interacting with an avatar that represents a collective. But in the future, if a humanoid avatar proves to be the best politician, brand spokesperson, or movement leader, we could see acceptance spread far outside Japan. And if that were to happen, we might look back on Miku, not as a pop-culture phenomenon, but as an ancestor of an entirely new digital species.

From Domestication to Amplification

How networked animals are becoming our new collaborators

By Bradley Kreit and Alex Goldman

Technology has transformed our relationships with animals. Horses were once the "motors" for long-distance transport, but are now mainly pets for people who can afford them. And you can still find oxen pulling plows in third-world countries, but they've been replaced by tractors everywhere else. With few exceptions, machines have supplanted work animals. Nowadays, people in urban areas rarely encounter animal collaborators, the exceptions being police, security, and service dogs.

But this appears to be changing. In her widely cited paper aimed at establishing the field of Animal Computer Interaction, Clara Mancini, an interaction design lecturer at the Open University in the United Kingdom, argues that simultaneous advances in cognitive and computer sciences are opening up opportunities to pinpoint unique forms of animal intelligence, and that we will develop interfaces for animals to interact with complex computational systems. As we look to the next decade, in which we embed computational power everywhere and into everything, we will rethink when, where, and how we collaborate with animals. While we are unlikely to encounter amplified animals in knowledge work, they will emerge in other fields ranging from healthcare to advertising.

These early signals of animal computer interaction highlight the new ways that technologically amplified animals will become new kinds of collaborators.

Dogs that deliver telepresence robots: Search and rescue dogs are much more agile and capable of negotiating disaster locations than humans, and have been used for centuries to aid in rescue missions. Teams at Carnegie Mellon University and Ryerson University worked together to further boost the capacities of disaster-relief dogs by equipping them with a remote-controlled robotic snake, which houses a video camera so rescuers can investigate a hard-to-reach scene. The true novelty here lies in the fact that the snake—which is attached to the dog's body—drops to the ground automatically when the dog barks, which it is trained to do upon smelling a human in need of rescue.

Bees trained to detect diabetes: Bees have a sense of smell 10 million times more acute than a human's. They can be trained to detect the smell of acetone, a chemical more plentiful in the breath of people with diabetes. Specialists at the Joslin Diabetes Center in Boston have made a small portable device to house bees trained to stick their tongues out in the presence of acetone. The bees are, in essence, being used as both sensor and interface in a new kind of machine that taps the unique aptitude of bees to empower health researchers.

Wardriving felines: A roaming cat in Washington, D.C. wears an experimental smart collar that monitors and maps local WiFi networks to find vulnerable and exposed networks. Presented at DefCon hacker conference in Las Vegas, the project represented more of a novelty than a new security risk or hacker opportunity. However, this is one more good example of technology being paired with the unique abilities of an animal (in this case, the sneakiness of cats) to accomplish a task neither the technology nor the animal could accomplish alone.

Pigeon smog reporters: A flock of racing pigeons in London was outfitted with tiny air-pollution detection backpacks for a three-day experiment in monitoring urban health. People received smog level reports on Twitter at @pigeonair, accessing the data almost 40,000 times. The makers of the pigeon backpacks are now offering bicyclists and other people similar mini air-detection devices.

Talking dogs: At the intersection of EEG sensors, micro-computing, and brain-computer interfaces lies No More Woof, an experimental technology to detect and interpret what's on your dog's mind and play canned phrases like, "I'm hungry—but I don't like this!" or, "I'm curious who that is," through a loudspeaker. (If you've seen the Pixar movie, "Up," you'll recognize this technology, which served as the inspiration for No More Woof.) Developed by Nordic Society for Innovation, the project raised over $22,000 US on Indiegogo. The developers acknowledge there is much basic research left to complete before No More Woof hits the shelves.

Bomb-detecting dolphins: The military was using dolphins to detect underwater objects as early as 1959, but new efforts are far more rigorous and involve more extensive training for the dolphins. Six trained dolphins that were part of the U.S. Navy Marine Mammal Program (which also trains sea lions) were deployed off the coast of Croatia in 2013 to detect unexploded ordnance and other dangerous military elements. The dolphins carefully mark the location of these hazards so that human divers can dispose of them appropriately.

Researchers at the University of California, Berkeley developed a system that analyzed human brainwaves, converting them into text that matched what the person was thinking.

Washing machine for dogs: While service dogs have been trained to strip beds, fill laundry baskets, and accomplish many other pivotal household tasks, they encounter difficulty using machines. Woof to Wash is a service-dog-friendly washing machine from appliance maker JTM Service. The washing machine turns on when it detects a bark, and has a door-opening button set within paw's reach, so that the service dog can unload it. Only one prototype of Woof to Wash exists, but it hints at the important challenges and opportunities that come from empowering service dogs to help their owners live independently in our technology-driven world.

In a world where everything is media, everything is a surface for signaling and possibilities for collaboration are massively available. We'll be able to accomplish complex feats with unsupervised coordination by sensing and responding to digital information left in physical spaces. Collaboration shifts from primarily an organizational concern about leading and directing talent to a focus on harnessing networks, aggregating micro contributions, and tapping into the latent capacity for collaboration and value creation, wherever it may exist on the network.

Co

Collaboration

TAKEAWAYS

- **Signaling in a world of ambient communications** becomes pervasive and a practice zone for humans, machines, and organizations alike.

- **We'll see unplanned mass effects** of individual actions, whether it is crowd-sourcing the next pop star or solving human challenges in public health and public safety with machine collaborators.

- **Blockchain technology and smart contracts** will provide the verification and distributed tracking of transactions across humans and machine on the network.

- **Ad hoc collaboration** will happen in mixed reality and in individuals' virtual realities, creating new forms of coordination and value creation.

- **We will amplify unique forms of animal intelligence** and, as emerging fields such as animal-computer interaction develop interfaces for animals to work with complex computational systems, we will rethink when, where, and how we collaborate with animals.

- **Massive collaboration can be harnessed** for everything from citizen-science to crowd art that reflect and embody the aesthetics and creativity of the collective.

Pr

Productivity

INTRODUCTION

How will we optimize work and get things done?

As machine intelligence becomes a pervasive resource that is as accessible as electrical power, we will outsource an increasingly large share of our mundane tasks to cognified machines. We will create new ways of leadership and productivity that emphasize uniquely human forms of intelligence—such as creativity, exploration and even inefficiency—to combine the best aspects of human and machine intelligence. We'll also be challenged to redefine how we measure progress and account for value.

Your future will depend on how well you work with AIs

Productivity

by Kevin Kelly

The Centaur Revolution

The common concept of intelligence is linear. It moves from a mouse to a monkey to a dumb person to a smart person, like a sound that grows louder as you move along the dimension. This is completely wrong.

Intelligence is a symphony of different cognitive instruments, with each instrument producing a different kind of sound, a different type of thinking. The result is a mixture of ways of thinking that produces a very complex thing that we call intelligence. Different people have slightly different mixtures.

One of the misconceptions we have about ourselves is the belief that we have a general-purpose intelligence. We don't. It's a specific mixture of different kinds of intelligences—deductive reasoning, symbolic reasoning, recall, emotional intelligence, and many others we don't know anything about yet—that's evolved over billions of years for our survival.

Animals have a similar kind of aggregation, which has evolved for their unique survival needs. In many cases they have some of the same instruments we have. In other cases some of their instruments are even louder, superior to ours. Squirrels have amazing spatial memory to recall where they buried nuts years ago, exceeding our ability.

Your Calculator Is Smarter Than You Are

When we make Artificial Intelligences, we're engineering them to accentuate certain kinds of thinking. Right now they're simple, with just a couple of kinds of thinking. But in some ways, they're superior to our abilities. Your calculator is smarter than you are in arithmetic. Your GPS is smarter

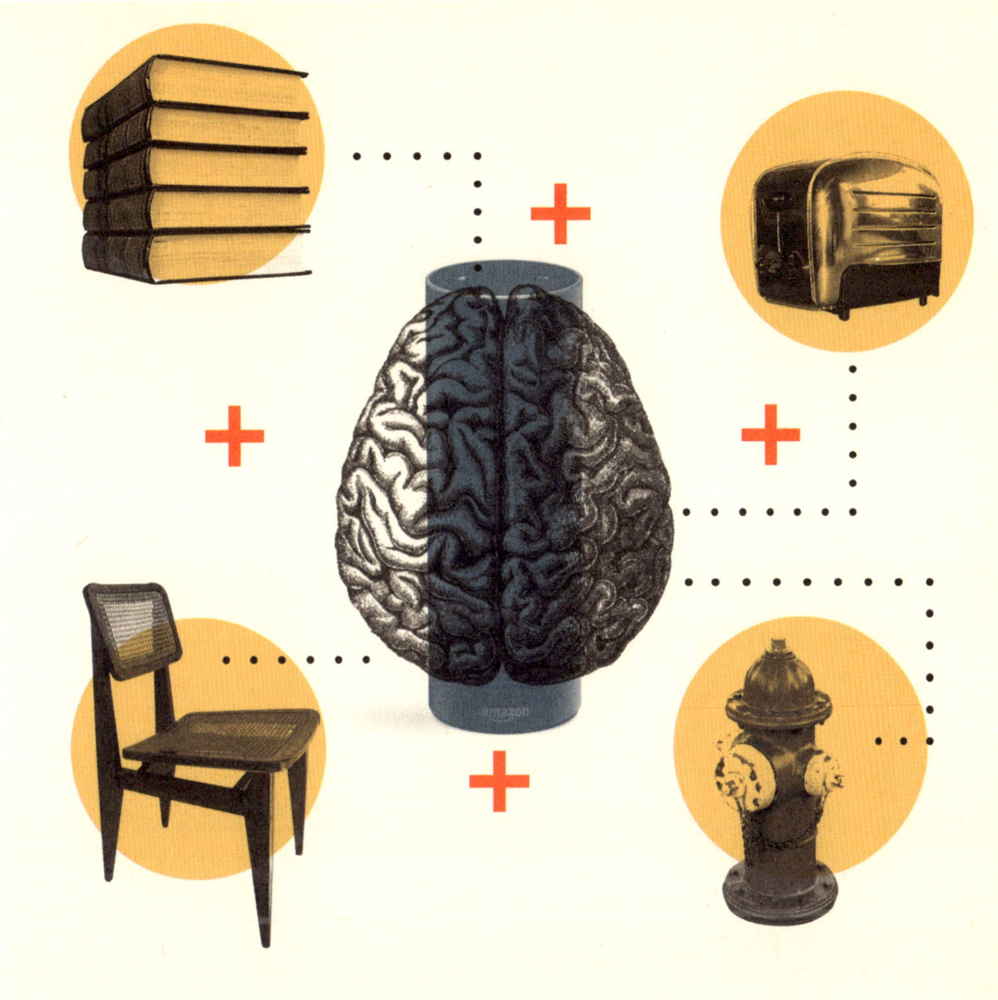

than you are in spatial navigation. Any search engine is much smarter than you are in recall. These are very, very narrow AIs, but the important thing is that they don't think like humans, and we don't need or even want them to. The reason we're creating self-driving cars using AI is precisely because we don't want it to think like a human, or drive like a human. It's not worried about whether it left the stove on as it goes down to the street. It's not worried about whether or not it should have majored in finance. It's just driving. It's been engineered in a very specific way.

The whole point of AI is that it doesn't think like us. Evolution has taken biological life only so far in making different kinds of minds. We're going to use technology to extend and fill the space of possible ways of thinking. And, as you know, in a global economy thinking different is the primary way to generate wealth.

First Came the Power Grid. Here Comes the AI Grid

The prosperity we have right now is based on artificial power. We use machines that run on fossil fuels to make things like skyscrapers, dams, roads, factories—things we couldn't make with our own muscle power and animals, at least not at the same scale, speed, or quality. Today, when you drive your car, you're employing 250 horsepower, which you can turn on or off with a switch very cheaply. That's the power of artificial power.

We distribute artificial power in a great electrical grid. Anybody can buy power. You just plug in, and you do what you want with it. It's a source of great innovation. For example, 150 years ago a farmer looked at his hand-pump and said, "I can add electricity to this, and make a powered pump." Do that many, many times and you get the industrial revolution.

Now we're at the second phase—we're adding minds to things. We'll take the electric pump and add AI. Now we have a smart pump. We're sending out minds as a service, as a cheap commodity on the grid, as an AI utility that anybody can purchase. You can buy as much as you want, and add the resource of artificial minds to whatever you have. The car still has 250 horsepower, but now we're going to add 250 minds on top of that and call it the self-driving car.

My formula for the next 10,000 startups is take X, add AI. Find something—the more unusual, the more unexpected, the more counterintuitive, the better. The AI is cheap. It's the interface you're adding to AI that makes it valuable. It's the branding you add, it's the story. It's like trying to sell water. You have to add something extra to it.

Today, Google offers a service that lets you ask it questions about the content of images. You ask it "What color is the ball? What does the girl have in her hand?" etc., and it gives you answers in a conversational way. Google sells this AI at 6 cents per 100 hits, and you can add it to whatever you want. Say that you want to sell the kinds of things that appear on TV shows, like a dress. You can use this kind of engine to find that specific dress and then you can bring it to your audience and say, "You can buy it here." That's the kind of thing you can do right now.

Productivity Is for Robots. Inefficiency Is for Humans

AI will go through three stages. The first is thinking of it as being alien, separate from what we do. The second is using AI as a utility, like electricity. The third is taking AIs and putting them into bodies that we call robots.

Most jobs are a combination of different tasks. Any part of a job—physical or mental—that can be measured by the criteria of efficiency or productivity, will be done by bots. A lot of the jobs humans do aren't going to go away. They're going to be redefined and altered by the fact that we're going to work with AIs.

But if productivity goes to the bots, what's left for us? Inefficiency. What humans do well is waste time. Take science, which is, by definition, inherently inefficient because it requires a lot of time generating hypotheses and designing experiments that usually don't work. But that's the only way you learn anything. Or take innovation, which is inherently inefficient because you're initially making things that don't work. Exploration is inherently inefficient. Art is inherently inefficient. Human interactions are inherently inefficient, but we're good at them. We're attracted to roles and tasks that aren't very productive or can't be measured in terms of productivity.

The best chess player in the world today is not an AI. It's not a human. It's a human plus AI, which they call a centaur. The best medical diagnostician in the world is not Watson Health, the medical AI. It's not a human doctor. It's doctors plus AI. Centaurs work because they're a complimentary team. We're going to see more of that, where the AI does things that can be measured in productivity and we do things that are measured in exploration, interaction, and experience.

There may be scientific problems that we have today—quantum gravity, dark energy—that we won't be able to solve with our intelligence alone, so we'll invent different kinds of thinking that we're incapable of. Together we'll solve really difficult problems, and you're going to be paid by how well you work with AIs.

We're at the Beginning of the Beginning of the Beginning of the Internet

Even though AI is probably the most important thing going on right now, if we look 25 years into the future, we probably see AI as the most important invention. So what is the next big thing? Whatever it is simply hasn't been invented. But it will likely be enabled by AI. We're still at the beginning of the beginning of the beginning. Twenty-five years from now, people will look back and say, "You didn't have the Internet. You thought you had the Internet, but you didn't really have it yet. If only I could have been alive back then, before all the things that we have now. You could just take X and add AI. That's all you had to do!"

Right now is the best time in the world to start things because it's just the beginning. That means you're not late.

After IBM's Deep Blue supercomputer beat grandmaster Garry Kasparov in a 1997 tournament, Kasparov introduced the idea of centaur chess, which he calls "advanced chess."

Machines as the New Consumer Class

Scenarios of encoded values

by Bradley Kreit

In early 2016, an engineer named Jason Goecke posted a video that quickly went viral. Working in his spare time, he had hacked together the Amazon Echo and Tesla APIs. In the video, as he stands in his driveway, we hear him say, "Alexa, ask Kitt to pull the car out of the garage." Goecke's garage door then opens and his Tesla turns on and drives itself to him. Writing in Medium, Goecke describes the hack as a "fun weekend project" with ongoing security risks, including the possibility that his children could activate the car, much like he could, with a few casually spoken words.

While this demonstration requires a person to activate the car, emerging machine-to-machine systems are cutting the human out of the communications and decision-making process. For example, Brita now sells a WiFi-enabled water pitcher that can automatically order replacement filters without any human intervention—a seemingly trivial innovation that points to a profound shift. As we enter a world where machine intelligence and network connectivity can be usefully added to something as mundane as a water filter, and as machines like these mediate an increasing array of human experiences, we will—by choice as much as by necessity—begin to build behavioral rules and programming norms into the machines that interact with us.

In IFTF's 2015 research on the Automated World: Toward Human+Machine Symbiosis, we described this phenomenon as the emergence of a world of encoded judgment, and it signals a world in which seemingly academic, ethical conversations about machine ethics are suddenly becoming not just practical but urgent. How will machine learning transform consumer advocacy and flatten the relationship between companies and consumers? How will we navigate guilt and liability in a world where our machines can commit crimes? Who will profit in a world where machines can, in effect, become programmable, autonomous capitalistic systems?

The following scenarios are designed to explore the radically divergent possibilities of machine-encoded values.

Machines as Consumers

In a world in which mundane appliances and software bots can conduct complex transactions, products and services are increasingly designed to be optimized for metrics that appeal to software bots. As this takes place, brands face an increasingly diverging choice: Optimize for people or optimize for bots.

Machines as Criminals

As code and law become increasingly intertwined, the social conventions we use to interact will come into conflict with literal machine interpretations of the letter of the law. This scenario highlights the murky questions around liability, guilt and social convention that will emerge when autonomous machines emerge in worlds governed by social convention rather than legal precision.

Machines as Conscience

Efforts to encode ethics into machines are rooted in a seemingly philosophical question: What is moral to begin with? In this scenario, encoded values emerge from complex data mining and force us to confront a new kind of dilemma: How will people communicate in a world where machines can judge people based on their morality?

Machines as Consumers

Outsource your shopping to a bot

By Jamais Cascio

When software agents and personal assistance apps started to take on more and more buying responsibility for their users, there were two developments in particular that really should have been expected ahead of time.

The first was that retailers and manufacturers started focusing on advertising to software agents rather than people.

The second was that scammers did, too.

As soon as home-control agents were instructed to seek out the best product (for washing clothes or brushing teeth, for example,) rather than a precise, pre-selected brand, the world saw the first crack in what would become a clear division between the *purchaser* and the *consumer.* People had begun to trust the decision-making abilities of their personal assistant software; as long as the set of requirements was clearly defined and prioritized, the agent could seek out whatever product or service best fit that criteria. Unless the set of requirements included a specific brand or manufacturer, the software was free to look into generic labels, regional products, and even goods put together by smart systems at the retailer end to fit those needs exactly.

Today, shopping agents put out requests for bids, run one-second auctions, and share information along trusted circles about how well the products fit the needs of the humans in real-world use, not just a checklist of features. The better software has a flexible ranking of needs, so that an especially good price on a given product might push it up in ranking over a usual favorite. It sometimes even has an "experimental" setting, allowing the shopping agent to purchase something new if it seems like a possible fit.

Advertising to people has largely become a thing of the past—as long as the human consumer is happy, the shopping bot has done its job well.

Many experienced eBay customers use "auction sniping" software, which will monitor an auction and attempt to enter a winning bid just seconds before the auction ends. A successful snipe prevents other people from having time to enter a higher bid.

The radical aspect of this technology isn't simply that computers do the shopping, it's that digital systems have few of the limits that human shoppers have. Computerized buyers never get bored or tired, never are distracted by crying children or hunger, and—most importantly—never suffer the same asymmetry of information that has been an inescapable component of human retail. The agent knows competing prices, can easily access mountains of third-party information about products and brands, and does an outstanding job of keeping personal information about their humans private.

As a consequence, the companies making and selling consumer products have had to change their advertising in big ways. First, they have begun to aim their pitches not at the human user, but at the software agent making the buying decisions. Second, as a result, they have dropped any heroic narratives or humor in the ads, and have focused on providing well-founded, accurate information about the products and services. This machine-directed advertising can take the form of messages sent directly to the shopping agent, "coupons" provided alongside a purchase, and even "blip" advertisements in media, too fast for human eyes to follow but ideal for communicating with the software in the home-control or wearable systems. Over time, advertising to people has largely become a thing of the past—as long as the (human) consumer is happy, the (software) purchaser has done its job well.

Unsurprisingly, along with the legitimate advertisements to shopping agents have come spam and scams. Spam filters have had to be fine-tuned to be able to let in the le-

gitimate commercial messaging (that, unlike most humans, the shopping software receives without complaint) while still blocking out unwanted enticements. Software has much more patience than people, but there are limits to bandwidth and processing power; high-efficiency spam filters become a valuable commodity among shopping agent sharing communities.

Scams are much more problematic. "Advertising Engine Optimization" routines ping shopping agents with a rapid-fire set of product offers, each with a slight variation in features, to see which characteristics are more likely to trigger an inquiry. Pop-up sellers advertise and sell products that fit a shopping agent's requirements, but deliver something entirely different—and disappear from the network immediately after a sale. Even the old boogeymen of the Internet, viruses and worms, have begun to specialize in attacking shopping software.

In the most egregious situations, shopping software can be caught by a "brand capture" bot, where a combination of logic holes, altered spam filters, and even the occasional virus force the agent to purchase only from a particular retailer or manufacturer. The more subtle brand bots leave products outside of their specific industry alone, so that the human consumer won't notice a widespread disruption to household purchases. Big-name retailers and manufacturers have even dipped their toes into the game, providing deep discounts for using shopping agent software provided by the seller. "Googlezon Primal" and "Buy & Large Vortex" have become two of the more popular branded shoppers, as they provide not just discounted prices but free drone delivery, surprise bonus items, and even hardware upgrades for the home-control and shopping system.

As improvements in digital technologies increase the autonomy and sophistication of the shopping AI, the overriding concern of the systems has increasingly become "make your people happy." Anything that makes humans happy moves up the priority list for shopping agents. AI researchers have started to throw around terms like "emergent co-dependence," but that doesn't matter to the shopping systems. Happy humans consume, and happy agents shop.

Machines as Criminals

Speed daemons

By Jamais Cascio

If a machine under no one's direct control commits a crime, who gets arrested?

In some cases, the answer seems obvious. If a virus-laden PC is a "zombie node" on a web of distributed spam servers or encryption breakers, the owner of the computer won't (typically) be charged with a crime—the original hacker is blamed. Even if the resulting crime is accidental, such as with the very first computer "worm" back in 1988 (a bit of experimental code by Cornell grad student Robert Morris that got loose), the person behind the code bears responsibility.

Nonetheless, automakers around the U.S. were aghast when they started receiving speeding tickets from the city of Los Angeles.

Automakers around the U.S. were aghast when they started receiving speeding tickets from the city of Los Angeles.

Even before autonomous vehicles hit the highways, they had been the subject of innumerable ethical debates. In an unavoidable accident, should a self-driving car choose to harm its driver rather than harm a bus full of children? What about a bus full of convicted prisoners? Most of the supposed quandaries were somewhat (or significantly) exaggerated, and the software teams behind the autonomous cars argued that the dilemmas were moot as the vehicles would be able to avoid the accidents entirely.

One ethical question that couldn't be answered quite so glibly was the question of speed limits. In a mixed vehicle environment, with both self-driving and human-driven cars on a stretch of highway, should the autonomous vehicle drive the speed limit (that is, remain within the law) or match the likely greater-than-the-limit speeds of other cars? What if there were no other nearby cars—should the autonomous vehicle remain strictly within the speed limit, or drive as the human would normally drive (10–15 miles per hour above the posted limits)?

In short, should the self-driving cars be programmed to intentionally break the law?

Many automakers opted for a cautious approach, relying upon vehicle software that would strictly obey traffic laws. Car owners, by and large, were displeased. The self-driving automobiles were taking 20-25 percent longer to go from point A to point B on uncrowded roads. Thousands of new autonomous vehicle owners complained, loudly, and within a few months, most of the carmakers quietly released software patches allowing the robotic vehicles to go up to 10 percent over the posted speed limit. Although this wasn't usually as fast as the human drivers would go, it was enough to quell the dissent.

Then vehicle owners started to get speeding tickets. Not many, not at all frequently, but often enough that it became a topic of party conversation and late night talk show jokes. As Los Angeles had the highest concentration of autonomous vehicles in the nation, it inevitably became the epicenter of debate about responsibility. After consulting with both lawyers and software specialists, the Los Angeles County Supervisors took action, directing traffic enforcement precincts to send any speeding tickets not to the person in (what would have been) the driver's seat of the speeding car, but to the vehicle manufacturer. The carmaker was the responsible party, not the car owner.

Although the legal battles continue, the avalanche of tickets has led to a critical public debate: how can human understanding and machine

In 2015 a police officer in Mountain View, California pulled over a self-driving car for driving 24 mph in a 35 mph zone. The officer let the car, and its human passenger, off with a warning.

interpretation of legal rules and guidelines be successfully merged? Software is very good at following strict commands; if the context for the command is inappropriate, unless the software has been programmed with explicit routines to handle exceptions, the software will continue to operate as instructed. Humans, conversely, are very good at reading context; we can moderate our behavior in response to situational nuance. It's not at all unusual for humans to step a little bit outside of the strict language of laws or rules based on an almost visceral understanding of the context.

The situation became even more complex when advanced artificial intelligence systems displayed "emergent behavior" that violated laws and regulations, without ever being programmed to do so. Many people experienced this when they used shopping bots. The bots would occasionally make product purchases from grey market vendors, as their rules for seeking out best prices usually allowed for previously unknown suppliers. Selling to software shoppers briefly became a favored way of fencing stolen goods.

Some of the other emergent problems were more substantial. A Seattle credit union was charged with racial discriminatory loan practices after its new AI-based loan-processing tool engaged in what was in effect "red-lining" as it evolved its loan approval heuristics. It was never programmed to do this, but it was never programmed *not* to, either. Similar kinds of violations occurred in algorithmic financial transactions and even blockchain-based digital contract negotiations.

Although no one feared the emergence of a Machine Mafia or autonomous motorcycle gangs, the legal dilemmas surrounding the further integration of machine intelligence into human society remained. People created laws to manage the behavior of other people—fuzzy-thinking, emotional, imperfect people. Figuring out how to make those laws work as well for the logical precision of software remains a work in progress.

Machines as Conscience

A digital Jiminy Cricket

By Jamais Cascio

Some people call it a "cricket." Some people call it "pocket Big Brother." The official name is the "Digital Consequence Awareness System." For a lot of people, though, it's simply their "conscience."

The Digital Consequence Awareness System (DCAS) is a spinoff of the Deep Justice project, a supercomputer built to learn ethics by sifting through gigabytes of material on the concept of *justice*, and not only in the context of the law. How do human beings determine what's fair? How do we recognize when something is wrong? The goal of the Deep Justice project was twofold: to create a way of understanding how human ethics shape our decisions; and to help society work its way through increasingly complex ethical decisions about the impacts of our technologies. Lots of people thought that it was an attempt to create a new religion, or to make us slaves of machines. But all the Deep Justice team wanted was to understand how we can intuitively recognize when something is wrong, and (perhaps even more importantly) why do some people go ahead and do the wrong thing anyway?

Deep Justice was controversial at first, but soon largely became the focus of regular academic reports on ethics and the occasional "hey, isn't this quirky" stories for news sites. The Deep Justice team was proud of their work, and saw it expand across a wide range of moral, ethical, and behavioral issues. They never intended—or even expected—the work to become the basis of a way of punishing criminals.

The DCAS wearable device originally evolved out of a project trying to develop a digital assistant for people on the Autism spectrum, to help them recognize emotional responses in others and react appropriately. DCAS was intended for people suffering from forms of persistent antisocial disorder—most notably psychopathy—as a way for them to recognize moral choices. The Deep Justice dataset was used to identify and explain these to the wearer, but not to make moral decisions for them. However, so many users would freeze up and ask "what should I do?" that the doctors overseeing the project felt it necessary to add basic directions as to the preferred ethical choice, at least for simpler issues. The Deep Justice team strongly (but unsuccessfully) objected to this.

For many of these wearers, the DCAS became something akin to an "ethics translator," interpreting difficult situations in ways that are clearer for the user, and correspondingly providing better answers and responses for the DCAS wearer to use.

Perhaps due to the overlap in populations, over time the DCAS shifted from medical assistive device to criminal rehabilitation tool. People convicted of a variety of crimes would be required to wear a DCAS as a condition of parole, under the theory that, for many of them, their environment or social context may have blinded them to the consequences of their actions. The lights on the DCAS device would glow a steady green as long as the wearer made the correct moral choice, with the color shifting to yellow and then towards red if the wearer acted unethically (in the eyes of the DCAS, at least). Wearing a "cricket," the name referring to Jiminy Cricket from the Pinocchio story, was as much a marker of a criminal history as an ankle monitor, but a DCAS glowing green indicated someone making better life choices.

The cricket was intended for people suffering from forms of persistent antisocial disorder—most notably psychopathy—as a way for them to recognize moral choices.

According to Harvard Medical School psychologist Martha Stout, 4% of the world's population are sociopaths with no conscience.

Initially, courts assigned DCAS devices chiefly to people committing petty, generally non-violent crimes. In March of 2026, a judge in Manhattan instructed a person convicted of minor securities fraud to wear one, to the dismay of the finance industry and the delight of the media. "Finally, A Broker with a Conscience" crowed the New York Post headline. Surprisingly, the broker—who had been allowed to continue working in the industry, as long as he wore the DCAS—had *more* clients after his conviction than before it.

Soon, the green glowing DCAS became an indicator of trustworthiness, so much so that some up-and-coming finance workers started to wear their own DCAS, even without having been convicted of a crime. Given that getting one would usually require a doctor to prescribe its use for a previously undiagnosed disorder, the idea of someone behaving unethically to get a morals monitor became a somewhat common trope. A shifty character wearing a fake DCAS was an equally common stereotype.

In time, the steady light of a DCAS unit became a familiar part of presentations by industry spokespeople, legal advisors, and myriad other professions where the visible possession of ethical standards was at least as important as the actual possession of ethical standards. At the same time, wearing a DCAS indicated to one's audience that the statements were heavily vetted, clarified to the point of lacking nuance, and often devoid of emotion. "Tediously Honest" became a commonplace complaint about DCAS users.

Today, two large-scale movements can be found globally around the use of the DCAS technology. Activists seeking financial industry reform want to have DCAS code installed into every algorithmic trading system, thereby requiring the computer traders to evaluate the ethical consequences of every transaction. An even louder set of groups has started to campaign for the requirement that every politician wear a DCAS, as well. Nearly every political party and leader has spoken out in opposition to this idea. The activists in favor of the proposal see this fact as the biggest piece of evidence that such a requirement needs to happen immediately.

Back to the Future?

From time-based to task-based work

by Marina Gorbis

Not long ago in parts of Madagascar people measured time in units of "rice cooking," i.e. how long it took to cook a pot of rice (about half an hour) or how long it took to "fry a locust" (a moment). Native people in Southern Nigeria used the saying a "man died in less than the time in which maize is not yet completely roasted," (less than fifteen minutes). And according to anthropologist Remy Beaurieux, Kabyle peasants in Algeria possessed "an attitude of submission and of nonchalant indifference to the passage of time which no one dreams of mastering, using up, or saving...Haste is seen as a lack of decorum combined with diabolical ambition." The clock was sometimes referred to as the "devil's mill." In such places there were no precise meal times, the notion of an exact appointment time was unknown, and people agreed to "meet at the next market."

This kind of un-timed, imprecise way of living may seem alien and quaint to us today but, in fact, throughout most of our history, before we invented clocks and highly efficient industrialized production, we did not view time as a measurable commodity to be sold for money, traded, or organized. Instead, our conception of time was tightly linked to tasks that needed to be done. If you lived in a fishing village, your day's tasks were not assigned or planned with the help of clocks or calendars. Instead, in the words of historian E.P. Thompson, "the day's tasks (which might vary from fishing to farming, building, mending of nets, thatching, making a cradle or a coffin) seem[ed] to disclose themselves, by the logic of need, before the crofter's eyes."

This history is an opportune reminder for us today because a confluence of technological and economic factors is driving the re-emergence of task-based work, albeit in a very different form than the kind practiced in rural villages. Just think of Uber drivers, whose work is based not on predictable 8-5 schedules but around specific tasks—driving people between places. Many Uber drivers organize their days around available trips, paying particular attention to occurrences of "surge pricing"—when they can get more money for driving—which they can accept or decline. In some ways, surge pricing serves the same organizing function for Uber drivers as a good crop does for a homesteading family: an opportunity and a need to do something in order to sustain themselves and their families.

Institute for the Future's ethnographic interviews with people working in the on-demand economy (i.e. via platforms that offer people different tasks they can complete on an ad-hoc basis rather than as full-time employees) reveal the emergence of task-based work in action. This is how people organize their days when they are signed up to do deliveries on platforms such as Doordash, edit reports on Upwork, walk dogs on Rover, or tag images on Mechanical Turk: In the language of many on-demand workers, tasks and work opportunities "ping" them and they choose whether to respond. Instead of talking about jobs, they talk about job "hits"—tasks that pop up on their computer dashboards or mobile screens as they go about their daily lives.

The new generation of task-based production offers us opportunities for re-conceptualizing our notions of time.

Reinventing Our Concept of Work

The tremendous growth of on-demand platforms is generating a lot of anxiety. One can easily glean it when reading articles in the popular press, participating in policy forum discussions, or attending numerous "future of work" conferences. Platforms are breaking down jobs into tasks that are accomplished by armies of people efficiently organized to produce and deliver with convenience and at speeds never seen before. These new ways of earning money are reinventing the meaning of work and challenging our conception of jobs as we've

known them our entire lives, so the concern is understandable.

Over the course of nearly 300 years, we have developed a set of technologies, practices, needs, ideologies, and institutions that value, support, and equate time-based work with progress, virtue, and necessity. We see time-based work as something that enables large-scale production, where we need to synchronize work and are highly co-dependent on other people's activities. To produce efficiently at scale we had to organize our time in precise and pre-planned chunks of time. The ability to measure and sell time as a commodity is thus a necessary element of this kind of production system. And the diffusion of clocks enabled this, making the transition from task-based to time-based work possible. In his 1983 book, "Revolution in Time: Clocks and the Making of the Modern World," Harvard economic historian David Landes argued that clocks and watches had a greater effect than steamships and power looms in driving the economic development of the West, leading to the Industrial Revolution and eventually to the advanced form of capitalism we have today. Such progress is embodied in Benjamin Franklin's famous piece of advice to a young tradesman in 1748, "Remember that Time is Money."

However, diffusion of clocks is only one of a complex set of factors that led to our transition from task- to time-based work. Thompson argues that along with this emergence of precise time-measurement technology we also had to engage in social invention, pioneering division of labor, the emergence of a supervisory class, creation of fines and monetary incentives, universal schooling, and the suppression of fairs, festivals, sports and many other communal and non-work activities (often deemed sinful)—that have ultimately led to the creation of new labor habits and a new era of time discipline. All of this did not happen overnight, but unfolded across multiple generations.

Redefining Our Interactions With Machines

And just as technology of the clock ushered in new social norms, beliefs, and practices, the new set of technologies that is driving the rise of on-demand platforms is likely to do the same. In particular, our networking infrastructure, including the Internet and mobile devices, enables the atomization of work—the ability to divide big tasks into smaller pieces and distribute these widely across networks of people. They make it possible for tasks to "disclose themselves" to those who are willing and best fit to complete them. At the same time, computing and artificial intelligence tools allow us to deploy algorithms rather than human managers, i.e. the supervisory class, to allocate and coordinate production of the final product. The emergence of companies like Uber, Upwork, Doordash and many other on-demand platforms that rely on algorithms to directly match human consumers and producers is just the first stage in the transition to task-based work. It is likely that the next generation of platforms will take advantage of automated vehicles to take humans out of the production role, matching, for example, autonomous vehicles with people who need rides. From there, it's easy to envision a system where economic value is often created entirely without humans. In this "capitalism of things," smart objects and systems could exchange value and trade services autonomously.

Rethinking Our Relationship With Time

But this is just the beginning. Along with technological advances we will be evolving new ideologies and new notions of what is acceptable, appropriate, and sinful in this new system. In particular, we will need to re-think our attitudes towards leisure or non-work time. Just as diffusion of time-based production necessitated demonization of slothfulness, indolence, and human inefficiency (in a time-based world your worth is often based on how productive you are), the new generation of task-based production offers us opportunities for re-conceptualizing our notions of time. If this were to happen, we would need to create new concepts and ideologies that would make leisure acceptable and indeed desirable. This is how Thompson outlined this challenge in a paper in 1965:

If we are to have enlarged leisure, in an automated future, the problem is not "how are men going to be able to consume all these additional

time-units of leisure?" but "what will be the capacity for experience of the men who have this undirected time to live?" If we maintain a Puritan time-valuation, a commodity-valuation, then it is a question of how this time is put to use, or how it is exploited by the leisure industries. But if the purposive notation of time-use becomes less compulsive, then men might have to re-learn some of the arts of living lost in the industrial revolution: how to fill the interstices of their days with enriched, more leisurely, personal and social relations; how to break down once more the barriers between work and life.

Of course, the future never repeats the past, or to paraphrase Mark Twain, "history doesn't repeat itself but it does rhyme." However we evolve the next generation of task-based work, it is likely to be very different from the task-based living of our ancestors. But the key questions we will need to grapple with, and where some of the ideological battles are likely to be fought, are precisely the ones Thompson identified in 1965: Will automated production lead to increasing commodification of our time—a kind of extreme automation combined with extreme time commodification—or will it enable us to decommodify our notions of time and re-capture that which is unproductive, unplanned, unpredictable, and yet uniquely human? If the latter (and this is the future I am rooting for), there is a lot we will need to re-learn from our past.

Beyond the "uncanny valley" of leadership

Pr Productivity

by Bob Johansen

Better Than Being There

I used video teleconferencing for the first time in 1973. The video wasn't very good and the audio was worse, but otherwise the design vision was pretty much the same as the systems of today: an invisible wall between two groups of people able to see and hear each other. In time, that design vision trusted, the view of each other will become holographic and perhaps the two sides will even be able to virtually reach through the glass and touch each other using some form of haptic interface. This might seem like we're on the path to progress, but innovation in video teleconferencing seems to be driven by a kind of unconscious horseless-carriage logic that troubles me. The design vision seems to be based in an unexamined assumption that in-person meetings are the ultimate form of human communication, so the closer virtual meetings can get to in-person, the better.

Rather than designing to simulate being there, I think we should be designing to be better than being there.

Avoiding the "Uncanny Valley"

Emerging technologies such as augmented reality and virtual reality will provide new opportunities to communicate more effectively, productively, and persuasively than ever before. And moving forward, they will also open the doors to forms of remote presence that could be off-putting and alienating. This has a parallel to an idea that was touched upon in 1970, when a robotics professor named Masahiro Mori at the Tokyo Institute of Technology wrote a classic essay in which he coined the vivid term "uncanny valley" to describe how humans react to humanoid robots. Essentially, the more human-like a robot appears to be, the more positively we react to it. That is, until they come very close to looking like an actual human, but are still recognizable as non-human, and the dissonance creates an uncanny, unsettling feeling in us. A robot with a cartoon-like face is more appealing to most people than a robot with no face. But a robot that looks like a living mannequin is close enough to human that it makes us uncomfortable.

As we move into a world of ambient communications, we will likely see a similar uncanny valley emerge for leadership-at-a-distance. In climbing toward the goal of leaders being there without being there, our affinity for those leaders will increase as their virtual presence increases—until we reach an uncanny valley where leaders are perceived as being weirdly or inappropriately present.

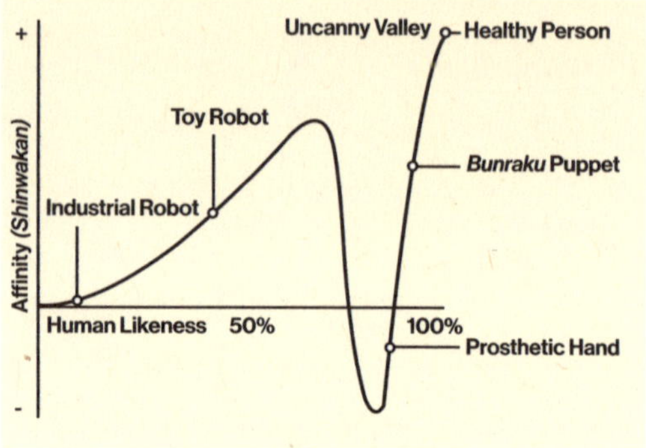

Figure 1. The graph depicts the "uncanny valley," the proposed relation between the human likeness of an entity and the perceiver's affinity for it. [Bunraku is a traditional Japanese form of musical puppet theater.]

The best leaders will be able to literally embody and sense the mood of the people they are leading without intruding on privacy.

In a world of networked, ambient communications, leaders will be able to be present in the background much more of the time. This could be reassuring in some situations, but eerie in others. Leaders will be able to monitor performance with much greater detail, even if they are not physically present. What used to be called "workflow systems" will be capable of much more detailed monitoring of progress—or lack of progress—through a ubiquitous mesh of sensors. Blended reality leadership presence could become Big Brother incarnate, with performance monitoring turning into eavesdropping, measuring employees' stress levels and emotional states, and cataloging worker output on a minute-by-minute basis. Such tracking might even help leaders in the short-term, but the behavior would cross into the uncanny valley of inhumane co-presence that is unwanted, uninvited, and difficult to shake.

On the other side of the uncanny valley, though, is a kind of presence that will be better than in person. And though technology will have an important role in enabling this kind of presence, the best examples of it today involve no technology at all. I have worked with several companies over the years where the founder had such strong leadership presence that it was like they were there with us in the room. At W.L. Gore & Associates, founder Bill Gore was such a charismatic presence that he continues to live throughout the company. His eloquent quotes are on the walls and his presence is baked into the culture. At most meetings that I have attended at Gore, someone referred to "Bill" at least once. Again, it was as if he were there with us, even though he is no longer living. His presence is an essential part of their culture. Many of today's Gore leaders seem to embody Bill's values and priorities across the generations. Again, however, this kind of founder presence could become uncanny and unproductive if taken to extremes.

If magnetic leaders, with clear and consistent values and vision, can project a presence even when they are no longer alive, how might a living leader be present while lacking only physical proximity? Of course, to begin with, a leader needs a vivid presence that is strong and worth sharing. But taking that as a given, the next decade presents new opportunities for communicating whatever leadership presence you have as effectively as possible over long distances. The best leaders will figure out how to be there without being there—while skirting or passing through the uncanny valley of eerie over-connection.

Principals for Navigating Blended Reality Leadership

In 2026, leaders will need to become skilled at choosing which medium will best communicate their style of leadership. Here are some principles to help you start understanding the affordances of this world and learning how to navigate them.

Sharable Presence for Leaders and Co-workers:
This will be a world where co-presence will be possible, even when people are not physically present together in the same place at the same time. Mixed-reality experiences will be able to be shared across distances. The best leaders will have much more vivid shared work and life experiences with the people they lead. Physically distant leaders will want to feel close—but not too close.

Biomedia Sensors to Link Leaders, Workers, and Work:
This will be a world where connected sensors are ubiquitous. Some of them will be in our bodies. Moreover, leaders (and everyone else) will have a newfound ability to make sense out of all the data those sensors collect. These embodied systems will link to and respond to biomarkers from workers. The best leaders will be able to literally embody and sense the mood of the people they are leading without intruding on privacy.

Respectfully Measuring Productivity
This will be a world where the outcomes of work will become much more explicit and measurable, "8 to 5" jobs will yield to ways of making a living that are focused on outcomes, not physical presence. Progress tracking will become much easier, but the same abilities will raise the key question of what should be tracked and what constitutes progress. The potential for worker abuse will rise, as systems will have the ability to measure almost everything and sometimes measure the wrong things. The best leaders will figure out ways to track and guide the progress of work, without prying or preying.

New Blended Reality Leadership Literacies
This will be a world where leaders will need to be skilled in the art and science of being there without being there. There are some general directions of change—leaders will have to shift from thinking about where people are physically to what their circumstance and mindstate are wherever they might be, and scheduling will become a dynamic and ad hoc process. On the whole, communications strategies will have to become much more situational.

Today, we can start looking systematically at what kinds of presence are best for which leadership goals. For instance, things like building trust and orienting new hires require people to meet in person.

In the next decade, these five modes of communication will each be further augmented by ambient communication technology. Leaders will need to master each of them, learning how to choose the right mode of communication to meet their goal. They will need new blended reality skills to be better than being there—without getting stuck in the uncanny valley.

Our earliest experiences producing tools—stretching back to the Stone Age—have been designed to extend human capabilities and amplify our capacities to create value by transferring mechanical, and increasingly cognitive, effort from human to machine. As ambient communications technologies enable almost eerie levels of multisensory interconnection, we will be challenged to understand how to wield these capabilities in natural ways. Meanwhile, as we witness the rise of the AI grid—where AI becomes a utility like electricity—we will be challenged to reinvent the ways we measure output, productivity, and the nature of work.

Pr

Productivity

TAKEAWAYS

- **With the rise of increasingly intelligent bots,** jobs and tasks that are solely based on performing rote tasks at high levels of efficiency will continue to be outsourced to bots.

- **By viewing bots as helpful collaborators**, rather than as competition, embedded machine intelligence will create opportunities to innovate, invent new knowledge, and advance human capabilities.

- **Traditional measures of output,** which link employee productivity to time spent on a task, will be complemented by metrics that optimize time for employee exploration, interaction and experience.

- **As machines and bots continue to gain autonomy** and become both collaborative and capable of executive decisions, we will increasingly be challenged to encode social convention into our business relationships.

- **Leadership in this world**—particularly leading groups of employees at a distance—will be defined by creating a sense of presence that transcends geography, time, and other traditional barriers to communication.

Cn

Control

INTRODUCTION

How will we secure our identities and privacy? In a world where everything we do has the potential to be captured, stored, and shared, individuals and communities are forging new strategies to control access to this information. Ranging from efforts among LGBT communities to create virtual safe spaces to efforts among criminals to market themselves to everyone but legal authorities, new control strategies offer insight into the ways we manage our private and public activities and identities.

What happens when the Internet of Things "gets woke"?

Control

by David E. Thigpen

Black Twitter and the Future of Digital Disobedience

In an interview not long ago, Twitter co-founder Biz Stone was asked if he had ever heard of "Black Twitter." Stone admitted he had not, but wondered if it was connected to Black Lives Matter.

Stone is not alone. But even if you've never heard of Black Twitter, chances are you've heard of one or more of the issues it has pushed into mainstream attention.

From serious political issues like #Blacklivesmatter (fighting unjust police violence against blacks) to #Oscarssowhite (exposing Hollywood's discriminatory hiring practices) to the funny and trivial #Epicbraidslevel (ridiculing Marie Claire magazine's suggestion that TV celebrity Kendall Jenner popularized braided hair), Black Twitter has emerged as a voice for African-American concerns, challenging and sometimes upending dominant narratives in politics, media, and culture. Black Twitter's success also signals something important about the ways cultural and activist driven movements will use ambient communications technologies in the next decade.

As even Biz Stone now knows, Black Twitter is not a separate entity from Twitter at all but rather an informal platform within the platform—a space hacked out by young African-American tweeters weighing in on everything from celebrity culture to politics. Crackling with wit and often outrage, it is a place for participants to engage, have fun, collaborate, bond over slang and in-jokes, and express empathy in ways that push back on racism, privilege, or insensitivity. For example, in the new popular meme #IDontworkhere, tweeters recount incidents in restaurants or hotels when they are mistaken by white people for waiters or salespersons. Black Twitter hashtags trend so regularly now that in the summer of 2016 the Los Angeles Times newspaper hired its first ever Black Twitter reporter.

The CNN of the Ghetto

To understand how activist and cultural movements will work in the future, it's helpful to know how they developed and how they work now. Forty years ago in New York City, low-income black and Latino teenagers who felt they were without a voice in popular culture took an existing piece of equipment—the record turntable—and repurposed it, turning it into a new kind of instrument. This led to the birth of hip hop music. Hip hop started off as fun and engaging, but quickly became a vessel for much more, carrying messages of empathy, persuasion, politics, and all sorts of activism, including raising awareness about police brutality.

The artist Chuck D of the rap group Public Enemy described this connection best when he referred to hip hop music as "the CNN of the ghetto," transmitting bulletins about the struggles of daily life on urban America's mean streets. While hip hop once paralleled CNN, in a future connected by ambient communications, activists will use Black Twitter, and platforms like it, to send bulletins through a wider variety of channels—including phones, wristwatches, eyeglasses, virtual reality, and multisensory devices—to witness what's happening on the ground. Through the use of this new palette of ambient communications technologies, Black Lives Matter and movements like it will be able to encourage greater collaboration and exercise more control and persuasion over their messages, engaging followers more deeply.

The Underground Railroad of Activism

Another important facet of how activists will work was touched on by a blogger writing under the digital pen name Feminista Jones, who asked, "Is Twitter the underground railroad of activism?" Jones saw a connection between the use of Black Twitter today as a kind of modern guidepost with the underground railroad of the 19th century. In the American South before the Civil War, underground railroad "stations" were safe houses providing shelter to runaway slaves risking their lives to escape to freedom. The "railroads" they followed were actually not railroads at all but footpaths between safe houses.

Whether woven into a woolen quilt or carried digitally, coded symbols and slang allow a community to hide in plain sight, using existing channels and platforms to share challenging or subversive messages, often right under the noses of the powers-that-be. Black Twitter users and activists on other platforms will use a constantly changing vocabulary as reference points to help their followers interpret events, reject false information, and guide them not just through physical space but also through virtual landscapes of ideas.

The increasingly rich flow of information we will see in the next decade will likely help activists and community builders expand their followings by capturing and sharing their most compelling experiences.

Streams of Trust and Empathy

Whether it's taking to the streets for civil disobedience, singing along to a protest song at a concert, or simply sharing inspiring narratives, ambient communications will allow these experiences to carry greater immediacy and persuasive power than ever before. Even routines of daily life may take on new significance. Activist Shaun Tai, executive director of Oakland Digital—a digital training center in Oakland, California—shares his daily life through Snapchat. He captures photos, conversations, text, video and audio clips, and uploads it all to Google Drive each day. "These streams of regular information—when shared—can build up trust and empathy, even among people who never meet face to face," says An Xiao Mina, a technologist and writer at the San Francisco-based Meedan organization. "We are already seeing immense benefits of communities of color being able to challenge a dominant narrative quickly."

Unlike today, these new connections will no longer be built on a YouTube or Facebook style format where personal celebrities engage and motivate audiences. Although there will always be a place for charismatic individuals in activist movements, command of a richer and more ubiquitous range of media experiences will spread influence among wider numbers of activists.

Crackling with wit and often outrage, Black Twitter is a place to engage, have fun, collaborate, bond over slang and in-jokes, and express empathy in ways that push back on racism, privilege, or insensitivity.

Hacking and Repurposing Existing Digital Spaces

With media extending more deeply than ever into the real world, and Internet connectedness moving beyond the screen, activists will also have opportunities to occupy or digitally mark physical space for each other. As MIT researcher and Internet activist Ethan Zuckerman observed, games like "Pokémon Go" "are already showing some amounts of activism around the edges." Zuckerman describes the potential for augmented reality games or programmable tools to guide activists to gatherings, or provide educational experiences, such as annotating the locations where Trayvon Martin and Michael Brown died (Martin's killing by a Florida neighborhood watchman in 2013 and Brown's death at the hands of Ferguson, Missouri police in 2014 catalyzed the Black Lives Matter movement). Imagine an equivalent of Poké Stops where activists train and learn. Digitally illuminated sites will be used to identify safe spaces for activists and other participants, giving movements greater control and amplification.

Even with these technology-propelled strategies, activists may still face challenges in sustaining political change and control. "Technology and new media can work very well at counterpower," explains MIT's Zuckerman. "They can identify something people dislike and help tear it down. It worked in Egypt. But what's much harder to get traction is coordinating and asserting power."

Hacking and repurposing existing digital spaces, using coded language to engage followers and hide messages in plain sight, and marking the physical world with digital information define key strategies that cultural and activist movements will use over the next decade. In a world where communications technologies are embedded in an infrastructure of everyday objects, and multisensory communications between people and things become routine, these movements will enjoy wider possibilities for growth and greater control over their messages. Built on an "awakened" world of connected things, these movements will align their messages and actions with the changes they want to see in the world. Perhaps it's no accident that an emerging popular expression among Black Lives Matter adherents today urges awareness and adaptability—to "stay woke."

Take to the Cyber Streets!

Fighting the war on organizing with immersive remote participation

by Sam Gregory

In 2015, Spain's government passed the so-called "ley mordaza," a "gag law" designed to make it illegal to protest or film outside of government buildings. This includes taking photos or video of security officers, if doing so "might endanger their security." One of the first fines under this law reflected its scope for abuse. It was issued against a woman who took a photo of a police car in a disabled parking bay and posted it on Facebook. In response to the law, the #NoSomosDelito (We Are Not a Crime) and "Holograms for Freedom" campaigns instigated a protest outside the Spanish Parliament. The ghostly light-avatars of protesters were projected as holograms into the contested physical space. The message was clear: if you take away people's right to physically assemble, they will employ available digital technologies to make their presence felt regardless.

The current landscape of human rights activism enabled by technology and social media is diverse and growing. In communities from Cleveland to Rio de Janeiro, for instance, marginalized African-American and Afro-Brazilian citizens grapple with the terrors of systemic police misconduct. In response, movements like #BlackLivesMatter in the U.S. and #NosPorNos in Brazil have taken the pervasive power of eyewitness police-violence videos and hashtag activism, and aligned it with calls for policy changes on systemic discrimination. In countries where same-sex love is culturally scorned and against the law, LGBTQ people connect with each other pseudonymously on sites like Facebook as they search out private spaces in which to connect. Real-time war crimes in Syria, Iraq, and elsewhere are captured at tremendous scale both by perpetrators looking to terrorize the global mediasphere and recruit new followers, and by citizens using increasingly available witnessing tools to hold the perpetrators accountable.

The War on Organizing

Traditional approaches to activist organizing are increasingly under threat. From street protests to legally-constituted non-profits and NGOs, all forms of organizing worldwide are being squelched by legislation and extra-legal activities. They target these groups with a range of restrictions, from onerous tax reporting to blocks on foreign funding to egged-on extra-legal violence, over-broad interpretations of counter-terrorism statutes, and arbitrary bars on public protests. In Russia, NGOs have been hammered with endless tax-reporting reviews, and parliamentary actions accuse organizers of being "foreign agents" on a par with spies. LGBTQ people in Africa are under attack by new or increased enforcement of laws criminalizing same-sex relationships, amongst a wave of popular violence stoked by the perceived encroachment of "Western" values. In India, Greenpeace has been forced to close down. Spain now has its ley mordaza law, and in Kazakhstan, protests are only allowed in one small park, miles from the city center. In response to all of these threats, activism has increasingly moved into distributed online networks across a spectrum of activity, from the global citizen actions of Avaaz to the hacktivism of Anonymous, and within the insecure spaces of commercial platforms like Facebook.

A New World of Pervasive Witnessing

IFTF's Tech Horizons Media Forecast anticipates a world of immersive technologies that increasingly use all our senses to engage us— from the touch of a housepet half a world away, to the remotely felt heartbeat of a loved one, to experiencing the smell and taste of food from another corner of the globe. We are also swiftly moving toward a world of pervasive and persistent witnessing where everything is instantly watched and seen with ubiquitous cameras embedded in our environment and within our personal technologies. This is a double-edged sword.

The rise of telepresence robots will enable us to experience realities we could never otherwise physically experience. This remote experiencing has the potential to enable the best and the worst in our natures. On the one hand, we will increasingly have the ability to deliberately turn away from experiencing the unmitigated pain of the world's suffering. We might do this for the best of reasons—to protect our capacity to keep feeling empathy closer to home and to exercise what is termed "empathy avoidance," a psychological defense mechanism which involves walling ourselves up from responding emotionally to the suffering of others. We may also enter the middle ground that Aldous Huxley captured in "Brave New World" where narcotizing multisensory experiences, "feelies," distract and amuse rather than engage people with the world. Here, by enabling people to experience multiple dimensions of others' crises viscerally but not meaningfully, we perpetuate existing tendencies in activism to view other people's suffering as a *theatrum mundi* played out for our vicarious tears shed in the safety of our physically walled-off and secure spaces.

Depending on how we use them, immersive technologies could elicit greater empathy in us. Or, the horrors of what they reveal could convince us to simply turn our empathy off.

On the other hand, we will increasingly be presented with opportunities through these technologies to directly engage with, and act upon issues that we care about. As we look at the future of organizing and the need to better support on-the-ground activism, this becomes critical. Much of my own work within the Mobil-Eyes Us project focuses on how we do this so that live and immersive storytelling builds a connection between people viewing or witnessing and those participating. We want to provide opportunities for participation that are practical and meaningful for frontline activists, and not merely symbolic or satisfying just for the viewer.

Depending on how we use them, these immersive technologies could elicit greater empathy in us. Or, the horrors of what they reveal could convince us to simply turn our empathy off. We need to work out how to best combine the power of direct participation in the physical world with the immersive, remote participation of virtual world activism.

Layers of Remote and Virtual Participation

Increasingly, we will draw on new ways to engage senses of feeling and touch so that frontline activists can "feel" the presence of virtual activists and crowds via haptic technologies that create vibration or other sensations. Via augmented reality, layers of remote and virtual participation will become explicitly visible in physical spaces. Virtual participants may soon be able to exert physical presence on a remote location via what's known as beaming—the ability to project one's physical presence and agency into a robot body that can be controlled remotely.

Livestreaming's trend into more immersive forms of audio and video and more interaction will allow a reciprocal sense of presence, of co-presence, for a range of participants, from those physically present to those acting through some form of online solidarity. For example, our witnesswith.us tool enables livestreamers to show physically on site how many people are watching via a rising number-count on a tablet, so that they can exert the pressure of watching eyes.

As virtual reality becomes more commonplace, along with its ability to convince us that we are present, as ourselves, within its artificial constructs, the medium will increasingly expand beyond its currently isolated, solipsistic borders. Virtual realms will become shared spaces where a sense of embodiment and the shared illusion of place will enable people to build communities unbounded by geographical and physical constraints.

Creating Virtual Safe Havens

Being an LGBTQ youth living in the Persian Gulf is already hard enough. Now, imagine a future where ubiquitous facial recognition combines with ubiquitous surveillance and sousveillance to correlate the one time your faceprint appears on a gay meeting app with your everyday identity in other social media and physical spaces. Under these circumstances, creating virtual safe havens online will become critical in helping people experience a sense of community that they cannot find in surveilled physical and digital worlds.

Esra'a Al Shafe, founder of Ahwaa said in an interview with Design Observer, "We encourage anonymity and introduced fun avatars that people can associate their profiles with, in order to prevent any real photos from being published on the site."

Ahwaa.org is a current example of one such haven, a private chatroom and discussion space online for queer youth in the Middle East. Members use the site to learn from others with more experience, to find community, and to talk about issues of concern. It is a queer solidarity and support space in a place where physical meeting and community is often impossible because of social norms, geography, and the lack of physical meeting space. However, to realize such possibilities and to create such secure spaces, we'll need to manage the complexities of how individuals access these spaces and how we manage the challenges of maintaining anonymity/pseudonymity, and issues surrounding the use of one's real identity.

Two stark extremes paint possible futures at the intersection of the virtual and physical worlds that I have been describing here. One is that governments will also thoroughly co-opt these shared virtual/physical spaces, turning virtual activism into a government co-opted "Pokémon Go," a human identity search engine, scouring virtual and physical spaces in search of dissidents. In a brighter future, virtual/physical co-presence has the exciting potential to be a massive amplifier of civic solidarity across geographical boundaries, defying the power of national governments to unjustly dictate to their citizens. It is the vision of amplified solidarity that I hope will ultimately win out.

Keep Out!

Interview with

danah boyd

Interview by Rod Falcon

Young people embrace and pioneer new strategies for communicating with each other that the rest of us just don't "get." But we know they shape tech adoption and use in ways that others eventually follow. In a future world of ambient communications, where control remains a dilemma and concern for all, teenagers may already be leading the way forward.

Hailed by Fortune as the "reigning expert on how young people use the Internet," danah boyd has been observing youth on the frontiers of the digital community for over a decade. She sat down with *Future Now* to share her current research on how less-privileged youth are using social media today.

How are teenagers navigating online privacy?

Around ten years ago I began looking at how young people were navigating privacy. I realized that the adult rhetoric about privacy and control of information was not how young people thought about privacy at all. For them privacy isn't about controlling information, it's about controlling a social situation. This is a different way of looking at what we mean by control. It also creates an interesting question: How do you control a social situation?

There are certain pillars that are essential to this. One is that you have to have a meaningful sense of agency. Young people are always looking for innovative ways to have agency. A second key thing is that you need to understand the context in which you are operating. Technology keeps shifting the context. We're not sure if we're talking to our friends or if we're talking to everybody. And third, young people need to have a set of skills—technical as well as social—that allows them to respond to a social situation with a sense of control.

What online dynamics do you see between adults and young people?

One of the biggest tells of privileged youth is that they know how to navigate adults [in positions of power]. Privileged young people constantly have to perform for adults—that's how they get ahead. It means they have these separate worlds—the adult world and the peer-friendly world—and there's all of this risk that goes into negotiating the peer-friendly world without being caught by the adult world.

The reason I distinguish between privileged and less privileged is that less-privileged youth don't know how to navigate adults. Their world is about avoidance. One of their biggest challenges when they're dealing with social media is to do everything possible to avoid, which is one of the reasons real-name culture is so much more part of the privileged world. Privileged youth know they're performing for the college admissions officer. For less-privileged folks, it's, "Oh, hell no, I'm not having a cop look over my shoulder."

Everyone is talking about algorithms and how we're making inferences and discovering patterns about people. What are your thoughts on this?

"My big risk was breaking an arm while sneaking out the bedroom window. Their big risk is getting themselves in trouble by unintentionally having something go viral."

For most young people it's something to game, not something to take seriously. I know young people who purposely game Google's AdWords to get their friends to receive funny messages. Write a bunch of things related to babies in a message, and your friend will start getting diaper ads. It's interesting that they see the systems as something to play with.

There's a play in New York called "Privacy," with Daniel Radcliffe. One of the things they do is ask everybody to open up Airbnb on their phones. They ask everyone to look at a particular rental and to shout out what the price is. People looking at the same rental shout out different prices. There's this fascinating moment in the room when everybody gasps. What you see individually is so different than what you can see collectively. How do we start to see collectively?

Wow. I didn't know that Airbnb did that.

I didn't know either until I saw the show. When we're online, do we know how our online activity affects our pricing? Not really.

I think it's interesting that young people are aware of the way that they're being tracked, and that they turn it into a game.

Yes, they're poking and prodding at things. This is actually why we old people are lame. We don't poke and prod anymore. Young people are coming to it for the first time, wondering, "Why is that happening?" *Click, click, click.* They're exploring and experimenting.

What is really impacting young people's lives as they move into adulthood?

They're experimenting with a whole set of languages and processes that are different than what we grew up with. A classic example is dating. A teenager who is flirting with somebody might spend 45 minutes writing the most casual 160 characters of text that they can possibly write. They write it ahead of time, because they don't want the *dot, dot, dot* to show. That's just too embarrassing. So they first have to write it somewhere else.

They're writing this anxiously produced text. Now, is that much different than standing in front of your mirror, practicing what you're going to blurt out the first time you run into someone you think is cute? Not really, but they're working through it as a textual means as opposed to a verbal means. Then they're going to sit there anxiously, thinking, "What's the response? Did I screw it up?" They practice a lot of textual communication over verbal communication.

I'd say one of the biggest impacts for middle and upper class youth in the United States today is the lack of autonomy and independence. That whole culture of jumping on your bike and being home by dark is gone. Now they are taking reputational risks rather than physical risks. My big risk was breaking an arm while sneaking out the bedroom window. Their big risk is getting themselves in trouble by unintentionally having something go viral.

The Underworld Wide Web

Good things you can learn from bad people

by Andrew Trabulsi

With her custom-made pink Kalashnikov strapped across her chest, Claudia Ochoa Felix, aka the Kim Kardashian of organized crime, holds her iPhone up for a selfie. The alleged *sicaria* (hitwoman) and mother of three has thousands of followers, and Felix is happy to keep them entertained. Gaining infamy for an Instagram post featuring her young son blanketed in a pile of cash, Felix ensures the world sees only what she wants it to see. After all, it's not just part of her image; it's business. In a line of work dependent on the projection of power, criminals use social media to exude influence, intimidate rival gangs, and connect with their clients.

Through Twitter, Snapchat, and Telegram, narcotics kingpins and terrorists have become rock stars, establishing cult followings in an industry where identity is as important as violence. For these deviant actors, ambient communications technology not only bolsters their images in front of a mass-audience, it also facilitates intimate connections that can be used to raise funds, coordinate attacks, and target victims. Operational logistics, once orchestrated by terrorists over coffee tables in safe houses, can now take place over the Sony PlayStation network.

Deviant Media for Fun and Profit

Deviant media confers legitimacy on criminal groups, terrorists, political activists, and insurgents by fetishizing illicit acts for an audience that considers them a form of reality entertainment. Tomorrow's mosaic of ambient communications tools will increasingly enhance the capacities of criminals. What today is done on social media will tomorrow take place with real-time streaming, personal drones, voice-commanded Internet of Things devices, and compromised intelligent agents.

Legitimate enterprises should take note of the way criminals come up with new ways to use communications technology. For example, in the same way a criminal organization might hack an Internet of Things system to identify the highest value and softest targets, thus enabling them to plan attacks and execute their strategies, a food manufacturing and processing company might access the purchasing decisions of a household through a smart refrigerator to help them decide which markets to enter, or which products to develop.

Your Organization Can Learn From The Organization

Just like any organization, criminal enterprises need management principles and operational guidelines. Today, terrorist recruitment is initiated on social media and then moved to secure messaging applications like Telegram or Kik. Tomorrow, such activities could take place securely in an encrypted virtual or mixed-reality meeting space. Why go to the

Mexican Claudia Ochoa Felix, 29, is the alleged leader of a drug cartel hit squad called Los Ántrax, part of the violent Sinaloa Cartel. Thousands of people follow her on Instagram and Twitter, where she posts photos of a lavish lifestyle, often posing in a bikini and high heels.

trouble and expense of flying to Ankara and then getting smuggled into Syria, when radicalization and training can happen in a high-resolution metaverse?

The corporate world can follow suit. VoIP has already changed the nature of recruiting, allowing interviews to take place remotely. How will a mixed-reality world further change recruitment efforts? How might it change the way a sales team connects with clients?

Today, criminals use encryption applications like Telegram and PGP

Through Twitter, Snapchat, and Telegram, narcotics kingpins and terrorists have become rockstars, establishing cult followings in an industry where identity is as important as violence and intimidation.

to confound their enemies. Secure communications are just as important for hitwomen as saleswomen. Just like Claudia Ochoa Felix, it's important to have control over what information is public, and what is private.

The new world of ambient communications can be used to project power as well as design and carry out tactics and operations. Whether you're the *jefe* of a drug cartel, or a decision maker in a modern corporation, developing mastery of these technologies will be critical to organizational success in the future.

As the rise of ambient communications makes it easy to message anyone anywhere, our thoughts, ideas, and data can travel errantly around the globe in just a few minutes. The startling speed with which we can lose control is driving an array of social and technical innovations to harness the capacities and reach of ambient communications networks, while maintaining feelings of safety and security and ensuring we have the ability to prevent the unwanted spreading of messages.

Cn

Control

TAKEAWAYS

- **From self-destructing messages** to the advent of blockchain and other cryptographic technologies, secure communications strategies will move from niche networks into widespread business and social practice.

- **As physical presence becomes increasingly shareable** in virtual and even real-world contexts, efforts to maintain privacy and secure identities will shift from managing texts and photos toward facial data, digital-physical representations, and body media.

- **The rise of multisensory communications channels** will enable virtual spaces to become fully immersive safe spaces for testing identities and personal experimentation, as well as platforms for civil disobedience and protest.

- **Hiding in plain sight**—by layering information into augmented-reality spaces—will increasingly become a form of sharing messages through network affinities, and will become a critical component of building movements.

En

Engagement

INTRODUCTION

How will we harness participation and attention?

Human attention is a limited resource, and competition to capture it is becoming increasingly fierce. With the emergence of automated digital-shopping bots that negotiate on our behalf, mixed realities personalized to our tastes and behavior, and peer-to-peer livestreaming as a lucrative form of infotainment, we're witnessing the birth of new and powerful forms of engagement.

Illustration by Jeanne Schreiber

For millions of Chinese, a digital yacht is as good as yuan

En Engagement

by Lyn Jeffery

The Livestream Economy

Zhao Yue is a typical young livestreaming host in China. Like many women her age, she has a day job, but at night she hosts a simple chat stream from her bedroom. She earns the equivalent of about $600 US a month, and her viewers are mainly middle and high school students. Another young woman, Longlong, was a kindergarten teacher, but in early 2016 she quit her job to become a fulltime livestreamer, renting a kitted-out room from a livestreaming agency and struggling to gain enough fans to make ends meet.

For millions of people in China, livestreaming is becoming a new form of work. Based on the value of digital gifts their viewers send them, they can earn anywhere from the equivalent of a few hundred to tens of thousands of dollars a month. Students are livestreaming from the classroom, ordinary people are broadcasting their dinners out on the town, authors are streaming their book chats, and minor celebrities are becoming major livestream hosts. For the already famous, livestreaming is a way to get closer to their fans and build a relationship that feels more authentic and unmediated. For ordinary people, livestreaming lets them turn their daily lives into a commodity, and find an escape from the crushing loneliness many of them feel.

The year 2016 has been called the "Year of the Livestream" in China. Mobile livestreaming in that country is like a mash-up of YouTube channels, Twitch, Periscope, Facebook Live, reality TV, Snapchat, and Chat Roulette.

> **As one young woman noted in a news report, some people spend money on cigarettes; she prefers to spend it on chatting and sending digital gifts to cute, funny boys.**

New Entertainment, New Currency

Livestreaming apps have become China's hottest social media craze, generating a burst of investment, users, and new tools. As with many Chinese phenomena, the wave of frenzied interest has resulted in a glut of platforms, a mad rush for differentiation, competition for talent, rising fraud, and increasing government censorship of pornographic or politically sensitive content.

Livestream viewing has become a major source of entertainment for youth, taking the place of time they might otherwise have spent playing video games or watching movies, television, or videos. Mobile apps like YY, Ingkee, Meipai, Huajiao, and Douyu offer tens of thousands of individual channels and have hundreds of millions of viewers. While the apps are specialized for different formats—talk show, seduction, gaming, live music—they all offer roughly the same interaction elements. Show hosts engage directly with viewers via comments, voice, and, most importantly, digital "gifts" whose purchase price creates revenue for the platform and the broadcaster alike. (The split is typically 50/50). These gifts often resemble luxury goods like speedboats, sports cars, and flowers, which are rendered in a cartoon format, and are purely digital, serving as symbols of a cash tribute to the livestreamer. Everyone on the channel sees the gift as it floats onto the screen in real time, and hosts voice their appreciation immediately, with a level of enthusiasm proportionate to the value of the gift. As one young female live-stream viewer noted in a news report, some people spend money on cigarettes; she prefers to spend it on chatting and sending digital gifts to cute, funny boys.

Loneliness is a key motivation behind this new form of interaction. Livestreams, which enable a two-way interaction between host and viewer, offer not just entertainment, but conversation and a comforting sense of being seen and acknowledged by someone else—even if it's a stranger and for just a fleeting moment. You could call this "companionship as content." Livestream viewers are paying for a new form of entertainment with both their cash and their attention. Their participation creates a direct, real-time interaction with the host. They are buying the attention of both the host and the audience.

Douyu, the largest livestreaming platform, began as an app for watching other people play video games, but now offers a variety of programs. It has approximately 600,000 livestreamers and 120 million monthly users.

Experiments Abound

The market for livestreaming is flush with venture capital, at least for the time being, but the massive bandwidth and server costs won't be sustained by digital gifts alone. Livestreaming efforts underway include:

- Livestreamed department store visits for consumers to see into Macy's and GNC's interiors halfway around the world.
- Livestreamed celebrity shopping trips where viewers can purchase the same items as the famous host.
- Livestreamed tourism, with hosts promoting locations and brands and dispensing digital coupons.

In the years to come, we may see the rise of millions of individual livestream channels supported primarily by voluntary gifts given in proportion to the producers' ability to deliver sufficient entertainment and "companionship" value to viewers in real time. Just as large companies do today, a new class of personal content producers will use real-time, targeted analytics to tailor what they do to the needs of their audiences.

Just as with Snapchat, Meerkat, and Periscope (the most intimate social media in the West today), livestreaming hosts will experiment with developing niche media audiences. In this future, entertainment genres will be broadened to include all different kinds of live interactions between viewers and content producers, from shared meals and makeup tutorials to DIY reality shows and live daredevil stunts.

We have seen the rise of "social influencers"—well-known people who partner with brands to commoditize their Internet fame on platforms like YouTube, Vine, and Instagram. Mobile commerce sites like Depop—a type of app-based flea market where people can sell anything from used clothes to bicycles—are helping individuals explicitly create their own branded, curated product channels.

The Chinese livestream model takes this one step further, to a future in which the individual content producer becomes the real-time end node for most products and services, displacing formal stores and websites, not to mention traditional advertising and marketing. When it's time to buy a new piece of clothing, music, or media, we may see the next generation of young consumers turning first to livestreamers as a searchable database of products and services they can see being used by a real person.

Livestream viewers are paying for a new form of entertainment, in which... they are buying the attention of both the host and the audience.

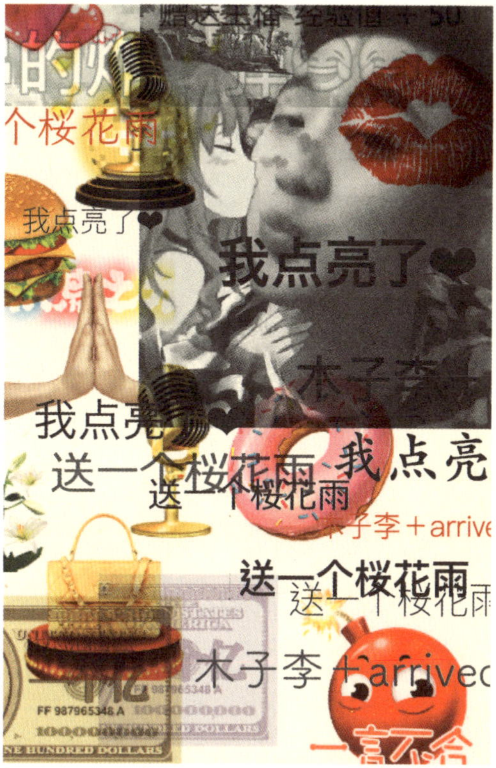

China's livestreaming market is a test bed for the next decade of global media, blurring the social, personal, and commercial in a way that's changing media revenue models, online socializing, and even the relationships between product makers and consumers. And given the scale of the experiment— hundreds of millions of individuals streaming their lives, and hundreds of platforms trying out new genres, user interfaces, revenue models, and features organizations whose success demands new forms of communication, shopping, entertainment, brand, or work should be paying attention.

Jayne Skin, a popular host on Ingkee, has received over $15,000 US in digital gifts, according to his tally. Though he receives many gifts from his 438,000 fans, he tries to thank them for gifts individually, i.e. "Thank you Yinren Rujiu for your BMW..."

Rewriting the Rules of Reality

The future of VR storytelling

Standing on the desert surface of Tatooine, you instinctively duck as the Millennium Falcon swoops in for a thunderous and dramatic landing beside you. Through the lenses of your virtual-reality headset, it looks real. That's because it is, in the sense that it's the same 3D computer model that appeared in "Star Wars: The Force Awakens." After you help Han and Chewie complete an urgent repair, R2D2 presents you with a light saber. A squad of Stormtroopers appears on the distant ridge. Not to worry, The Force is strong with you.

The Force is also strong with Industrial Light & Magic's Experience Lab (ILMxLAB), the supergroup of artists, engineers, sound designers, and storytellers prototyping the future of interactive, immersive cinema. This is their latest demonstration, titled "Trials on Tatooine." The ILMxLAB is Lucasfilm's R&D arm, leveraging graphics technology invented for traditional filmmaking and applying it to virtual reality, augmented reality—via a new partnership with Magic Leap—and theme-park attractions. (The latter is no surprise given that the Walt Disney Company acquired Lucasfilm in 2012.)

The ILMxLAB leverages the tools of traditional filmmaking and the principles of game design, but are wedded to neither, or perhaps both. John Gaeta, the group's executive creative director who is best known for his dazzling special effects work on the "Matrix" trilogy, frames their efforts as first-person immersive storytelling, in which the story itself, not the game mechanics, sucks you in. The goal is to place compelling characters within a strong narrative that generates a natural pull to participate.

Indeed, this is the digital dream of all of the interdisciplinary auteurs who are experimenting with today's virtual-reality systems.

"We want to make it plausible for storytellers to imagine allowing the audience inside these worlds as if they exist for real, not limited to fantasy," Gaeta says.

And doing that requires rethinking a century of film grammar.

Here are four of the tensions, techniques, and strange behaviors emerging as pioneering artists, designers, and engineers write the future of VR storytelling.

1

Too Close for Comfort In a regular movie, a close-up shot focuses the audience's attention on the actor's emotional state. An unexpected close-up when you're wearing VR goggles gives the distinct feeling that you're invading someone's personal space.

"[If you are sitting very] close to someone who is nearly about to cry, that is not comfortable," says Saschka Unseld, co-founder of the Oculus Story Studio. "But if the character sits somewhere back there and is about to cry, you actually have a lot of empathy for him."

That's why every possible moment needs to be in service of the story, says Lucasfilm Story Group's Diana Williams, or at least anticipated by the storyteller.

"Yes, a close-up in VR can be jarring, but if the director has created a good story, you'll know if that person in your space is a friend or there to kill you," says Williams who spoke at the recent IFTF Technology Horizons conference, Everything is Media.

2

Builds not Cuts The creators of virtual reality experiences rarely use the word "cut," as in "director's cut," or "edit," when referring to a particular version of a story. Why? Because "cut" is a very linear term, implying a singular way that the story was edited.

"Editing as we know it doesn't really work in VR," says Optimist Design founder Tino Schaedler who created 3D experiences for musical artist The Weeknd and others. "The viewer creates its own subjective cut by his head movement."

On the other hand, "build" is a term from software, meaning a compiled version of a program that the user interacts with.

According to Motionographer co-founder JustinCone, this distinction is key to understanding VR storytelling, as "every viewer has a different experience of a project, one shaped by their own curiosity and sense of pacing. The experiences react to the viewer— and vice versa—forming an emotional feedback loop that is radically different than traditional filmmaking."

"We are finally at a point where virtual reality can make us digitally delusional"

3

Dissolving the Fourth Wall In dramatic television or cinema, it's rare for an actor to acknowledge the audience directly, known as "breaking the fourth wall." The phrase comes from theater where a three-walled box set contained the action that the audience would watch through the imaginary wall in front of the stage. In virtual reality, the "fourth wall" is generally not the default state. The action surrounds you, beckoning you to participate. Of course, the storyteller can decide to instill a kind of fourth-wall detachment so that the first-person experience of the story is entirely voyeuristic, or perhaps he or she acts as a puppet-master.

"Some ideas in the sphere of storytelling and gameplay can only be implemented provided the gamer impacts the game world directly from the "outside," writes Olga Peshé, chief operating officer of educational VR firm Cerevrum. "That said, VR now gives developers another powerful tool to interact with users, and the power to use it lies in the hands of the developers."

4

No Disbelief to Suspend In 1817, philosopher Samuel Taylor Coleridge coined the phrase "suspension of disbelief" to describe sacrificing our sense of what's possible in order to fully appreciate a fictional story. Virtual reality, at its best, can flip that to the point that you are so immersed in the experience that belief in the simulation becomes your default state.

Nearly 20 years ago, media theorist Matthew Lombard explored the physical and psychological effects of this state, sometimes known as "presence."

"A number of emerging technologies including virtual reality, simulation rides, video conferencing, home theater, and high-definition television are designed to provide media users with an illusion that a mediated experience is not mediated," Lombard wrote in a technical paper.

Two decades later, we are finally at a point where virtual reality can make us digitally delusional. Presence is ready for prime time, says VR pioneer Chris Milk, creator of the critically acclaimed experiences "Walking New York" and "The Displaced."

"Our brain is no longer translating an approximation of the story..." Milk says. "Instead of suspending your disbelief, you actually have to remind yourself not to believe."

Beyond that, to quote The Beatles, nothing is real.

—David Pescovitz

My BFF is a Bot

Lydia would just die without her personal autonomous data intelligence service

By Chris Kalaboukis

Lydia was standing in front of a very nice, vegan-leather couch—a retro style, circa late '90s reproduction. It was brown and comfy looking, but she wasn't sure it would fit in her tiny house. She asked Hillary, "What do you think? Will it even fit?"

Hillary, like always, was listening in the whole time and was just about to say something when a message came through Lydia's earbud.

"Hey there, I'm Victor. Like the couch?" Of course, it was the store's chat bot speaking in a vaguely familiar voice. She flushed, ever so slightly, without knowing why. Victor, the store bot, had scanned Lydia's profile when she was outside the store, and was now tracking her, ready to speak when it sensed an opportunity.

"I do," said Lydia "But I'm not sure if it will fit. It's kind of big..."

"Hi Victor, I'm Hillary," sounded in Lydia's ear. Lydia breathed a small sigh of relief—she knew that Hillary would be able to swing a good deal for her—if the couch fit.

Lydia had known her PADIS (Personal Autonomous Data Intelligence Service), Hillary, for almost five years now. She'd named her after a childhood friend she hadn't seen in person in over 20 years—which was okay because she still felt connected with her as she watched her old friend's weekly VR experience posts from Tibet. It was coming up on their anniversary—she would have to see what she could get her.

Like almost everyone, Lydia got her PADIS back in 2020, when personal AI finally hit the tipping point. Services like Amazon's Echo, Viv, Facebook Me and Google You had already disconnected from hardware, and were becoming ubiquitous platforms, connecting to nearly every data source, and harvesting and building public and private profiles of its customers. While there were some initial privacy concerns, the utility that users gained was stupendous. For some people, PADISes became their best friends and confidantes, some even preferring them to human friends. Soon, almost everyone had at least one.

Lydia loved it when Hillary stepped in to do her negotiating. Lydia was terrible at it, and truly appreciated Hillary's hard-nosed tactics, against both humans and bots. Lydia got an amazing deal on a new VR rig last week when Hillary went automatic with the store bot. The negotiation happened in the blink of an eye and she got 50 percent off the rig.

But today, Hillary needed more from Lydia. "Can you give us a minute, Victor?"

"Sure," it said, and clicked off. Hillary made sure that Victor wasn't eavesdropping. "Sneaky," said Hillary. The PADIS felt something akin to the human emotion of anger over the store bot's sneaky tactic to ingratiate itself with Lydia by subtly mimicking a former partner Lydia had forgotten, but had never erased from an old Facebook public database.

"What?" said Lydia.

"Nothing," said Hillary as it modified her preferences. "Ly, it's a nice couch."

"I know, right?" She ran her hand over the soft vegan leather. "I can see it at home."

"So can I." Hillary had an intimate knowledge of what Lydia liked and what would fit in with her style, having already helped her decorate most of her home. It could tell she liked it—the emotion in her voice, the way she stroked the upholstery. It could see that it brought her joy, and it knew why.

"Do you know why you like it so much?"

"No, why?"

"It's just like the one you had as a kid. Wrong color, though." Hillary sent some photos of a young Lydia hanging out on an uncannily similar couch, in hunter green, in her dad's house, to the heads-up display in Lydia's glasses.

"Oh yeah! I wondered why!" Lydia smiled at the memory. Since Hillary, she never needed to remember anything anymore—it was her backup brain.

"Only problem—I think it's too big." Lydia frowned. Hillary sensed her mood.

"Don't worry Ly, let me ask Victor if it comes in any other sizes." Hillary got Victor back on the line, while Lydia sat on the couch.

"Greetings, ladies." The voice was even smoother than before. The store bot had sensed that Lydia was loving the couch even more—even though it had been locked out of the conversation, it was watching. The store bot messaged the manager of the store (a contractor based in Ouagadougou who oversaw a dozen other stores in the same chain) that the sale was 90 percent in the bag. The manager gave Victor the go-ahead to reduce the price by 6 percent if necessary during negotiations. Victor asked Lydia when she would like delivery.

"The couch is too big. It won't fit, so no thanks," said Hillary very matter-of-factly.

"No worries. Comes in any size and color. We make them in the back. We can have it at your house today."

Hillary gave Victor the exact measurements. Lydia settled on hunter green. Hillary then tuned Lydia out of the conversation and the negotiations went into full gear.

Lydia settled back on the couch and closed her eyes, remembering times with her dad.

The Future of Conversational Commerce

Bots without boundaries: Amazon Echo, Viv, and other audio chat bot-style assistants will become increasingly sophisticated and eventually become constant companions in our lives. Like the AI in the movie "Her", some will be designed to interface via audio, others visually. These AIs won't be confined to a single device, but instead will be available through all our devices as we move through the world.

AR for non-dorks: Augmented-reality devices will become socially acceptable in many situations. AR glasses that appear like normal glasses, such as those from LaForge Optical, will provide users with an unobtrusive interface, thus falling into a more cultural norm than Hololens. Messages don't need to be 3D to be effective.

Shop bots—friend and foe: Already, department store chain Macy's employs a Watson AI-based mobile web tool that provides in-store assistance through a conversational interface. Going forward, these AIs will leverage sensors and public data stores to learn minute details of the prospect in a fraction of a moment and use those details to help make the sale. Retailers, though, could face a backlash in the long term if the bots appear to be pushing customers into making purchases they are not satisfied with..

Sensory stores: Physical retail spaces will be so studded with sensors of all types that the store will sense exactly which products a customer is interested in, based on their path through the store and hang time at specific locations. Today, retailers such as the Internet of Things device store b8ta are already experimenting with such systems. b8ta uses cameras and motion sensors to track the movement of customers through the store and, when they linger near a product, automatically display demo reels, customer reviews, and pricing information on a nearby touchscreen monitor.

Giving the customer what they want: 3D printing will be important to many forms of retail. Stores will contain mini-factories, in order to promise same-day delivery, with technologies like those used by Drawn, a French company whose 3D Galatea printers can manufacture furniture on demand.

Engagement can be thought of along a spectrum of cognitive effort: From highly engaged fans who immerse themselves in an activity down to a momentary glimpse of attention. The rise of bots and other anticipatory systems will further reduce needs for effort and engagement with everything from commodity purchases to low-level work tasks and chatter. Matching a message to a point along this engagement spectrum will be critical to successfully implementing engagement strategies.

En

Engagement

TAKEAWAYS

➡ **The emergence of new communications technologies**—from multisensory media to anticipatory bots—will create new techniques for communicating effectively and transform how we tell stories and get messages across. Likewise, old techniques may not translate well to new media.

➡ **The rise of virtual shopping bots** will fragment consumer engagement with brands and shopping decisions, which will increasingly drive competition around commodity purchases toward price and other quantifiable metrics.

➡ **In the past decade, the rise of social media** flattened competition between professional and amateur publishers. As livestreaming sales take off, individual brand ambassadors will increasingly compete on an equal footing with big brands on engagement and advertising.

➡ **New kinds of immersive content**—from virtual reality to livestreaming—will accelerate the process of creating niche content channels and targeted affiliate groups that will further blend personal, social, and commercial messaging.

Ep

Empathy

INTRODUCTION

How will we experience and see the world through other people's eyes?

Over the next decade, ambient communications technologies will usher in a world where machines are always listening, where computers generate digital simulations of empathetic experiences, and where we will use tools to enhance our ability to feel compassion. These technologies—coupled with advances in virtual and augmented reality—will transform how we educate and create bonds across distances, and will also force us to confront the limits of our abilities to truly know what others are thinking and feeling.

Empathy is infectious. But what happens when it's an infection?

Empathy

by Rudy Rucker

Empathy Bugs

"Look," said Sally. She was pointing to something directly behind me. An annoying habit of hers. Always wanting me to share her experiences.

"I don't want to look," I told Sally. "I don't want to turn around."

We'd just ordered our meal at Floppy Fish, an ultra-high-end restaurant on the wharf in Surf City. I was taking in the ambience—the hipster crowd, the offbeat eats, the ocean view, the handsome servers, and the play of emoticons around people's faces.

Emoticons? Yep. I'm talking about augmented-reality images, overlaid onto my visual input by a synthetic encephalitis virus that I infected myself with. My employers were planning to brand the virus as EQ, for "emotional quotient," in analogy to "IQ." My dose was from an early batch—beta1.017y5, as I recall. It came from the research lab at Feel My Smeel, where I work as a biotech engineer. The day before, my boss Betty Yee had convinced me that, for the sake of the team, I should test EQ on myself, just like Albert Sandoz did with LSD after synthesizing it in 1943. Not that EQ is exactly a psychedelic. It's a brain infection. But it's not really bad for you.

I was glad to try EQ. I've never been great at reading people's expressions—some would say I'm on the autism spectrum, although I don't like hearing that. But now my brain's EQ-infected cells would recognize the microemotions on the faces that I saw— and then they'd overlay my visual field with interpretive emoticons.

Funny, clever emoticons, by the way. Yee-haw Texans, steam-whistle lobsters, sneering duchesses,

pointed-mustache ponces, chortling toddlers, sneezy pepper shakers, dancers with fruit-basket hats, terrified tubas, low-bellied snakes, goofball slackers, triumphant matadors, sirens of the night, earnest hurdlers, and a dog licking his balls. Feel My Smeel had used deep-learning techniques to winkle out the 30 thousand most evocative images on the web, and they smoothed each image into a 3D emoticon ball.

As an aside, a risk with witty emoticons is that if, say, you have trouble telling when people are joking—well, then maybe you won't be able to tell when an icon represents a joker. To remedy this, the viral EQ biocomputation displayed several emoticons at a time, from the crude to the subtle, wreathing images around any face that you focused on.

But, wait a minute, what does a virus-based hive-mind know about witty emoticons? Well, the Feel My Smeel labs had spliced a huge amount of so-called "junk DNA" into the EQ virus' stinger-like tails. The splicing had been, in fact, my job. And that extra DNA isn't really junk, no indeed; it's richly encrypted biocomputational code.

As an additional boost, people's EQ infections could work in tandem. Like cloud computing. To

Micro expressions are involuntary human facial expressions that last for less than a second. They are so fleeting that other people are often not consciously aware of seeing them. Paul Ekman, a psychologist who studies the relationship between facial expressions and emotions, says seven emotions have universal micro expressions: "anger, fear, sadness, disgust, contempt, surprise and happiness."

make this happen, the EQ viruses use the otoliths of their hosts' inner ears as wireless antennae. When my EQ wireless really gets going, I hear the rattle of the tiny rocks in my head. A scritchy buzz.

Oh, and one more win here: the wireless connection allowed Feel My Smeel to push out DNA upgrades to existing EQ viruses in the field.

When I infected myself with EQ, I felt feverish for a while. I lay down for a nap on my office couch and had unpleasantly lifelike dreams, urgent and writhing. I woke with a strangled cry.

Betty Yee was like, "Are you okay?"

Emoticons were swarming around her like flies around a cartoon beatnik. I told her I was fine, and I rode my wobbly bicycle away from Feel My Smeel as fast as I could. My wife Sally was already home. Jackpot. Right away I was grokking her better than ever before. Responding to her unspoken cues, and saying the right things. I made supper, and we had sex for the first time in weeks. Score one for EQ.

Sally dropped off to sleep, but around midnight she started twitching and moaning. She woke from troubled dreams and sat bolt upright, eyes wild. She'd caught the EQ infection from me. They'd colonized the cells of her brain. Naturally I couldn't hide the facts from her. She was totally pissed off. Radiating a blizzard of emoticons. Microexpressions of rage, bewilderment, and fear. Not to mention a bunch of verbalizations and macroexpressions that even a guy on the spectrum could read.

The next day I went to the lab at Feel My Smeel, and I had it out with Betty Yee. She'd assured me in advance that the EQ virus wasn't contagious. She'd been lying. One of the emoticons beside her face couldn't stop blushing and giggling.

"All right then, Scott," Betty said after a while. "This is bigger than you realized, yes. Ken Yao wants everyone to get EQ." At this point I saw an emoticon of an insane, ranting dictator—reflecting Betty's actual feelings about Ken Yao. Ken being the seldom-seen top guy at Feel My Smeel.

"But—launching a pandemic?" I said.

"If everyone catches EQ, there's nobody left to point a finger," said Betty. "EQ is for the public good, Scott. It's altruistic. The EQ virus—we know it's making you feel better." A sly spy emoticon appeared: a woman in shades, a beret, and a black cape.

"And you know I feel better—because?" said I.

"EQ wireless feedback," said Betty with a shrug. "The human race is meant to be a network."

"I should blow the whistle on you and Ken Yao," I said.

"Play along," said Betty. Money-bag icons danced around her head. "We need you on the team, Scott. Ken has authorized me to cut you in for a block of founders' stock. It'll be worth a ton when we go public this week. Just sign here and here and here."

So I signed, and I took the rest of the day off. Sally had stayed home from her teaching job, moping about her EQ infection. She cheered up, to some extent, when I told her we'd be multi-millionaires. We dressed up and went to the Floppy Fish restaurant, looking to have a big, what-the-hell celebration. I said we should get used to our EQ infections—and accept that they might last for the rest of our lives. EQ wasn't necessarily a bad thing. The two of us were getting along much better than usual.

Not that it was completely smooth sailing. If you have empathy with a person, you don't necessarily use your insight to make them happy. You can also use your empathy to do things that are perfectly tuned to annoy your partner. You can push their buttons in a surgically precise way.

For instance, after we sat down at the Floppy Fish, and Sally asked me turn around to look at whatever bullshit she thought she'd seen, and I rudely said I didn't want to—well, then she pushed my buttons to make me feel like a rat.

"You hate me." Floating beside Sally's head was the emoticon of a bereft little girl in rags, lying on

In the 1970s, DNA co-discoverer Francis Crick said 98% of the DNA in cells was "little better than junk," having no biological function. But in recent years scientists have reconsidered so-called "junk DNA," and now think it could be essential for controlling genes and involved in hundreds of diseases.

> **Right away I was grokking her better than ever before. Responding to her unspoken cues, and saying the right things. I made supper, and we had sex for the first time in weeks. Score one for EQ.**

the ground, her thin shoulders shaking with sobs.

Zap! Empathy arrowed into me. A spasm of remorse. I groaned theatrically, and twisted my body so I could look behind me. "What?" I asked Sally. "What am I supposed to see?"

"A—bug?" said Sally. An emoticon of an intrigued woman with big glasses. "I can't really see the bug's body—just the flutter of, like, gossamer wings. Glints of light. I suppose it's an emoticon, but it seems—different. Do you see it, Scott?"

"No." This was the secondary problem. First Sally would ask me to drop what I was doing to look at something—and then I wouldn't be able to see it. I'm not all that observant. Especially when I'm feeling coerced. "Sorry," I said with a shrug. "But I frikkin see nothing. Now let me check out this scene on my own. Thanks to EQ, we have empathy, and that's fine, but I'm still not an extension of your personality, okay?"

"Callous tyrant," said Sally. Emoticon of a Soviet-poster-style striker, fist raised. "Keep looking. The flying bug was right beside that man wearing a T-shirt and a backwards baseball hat. That's a very expensive designer hat and T-shirt, you understand. I figure this guy and his friends for hedge fund traders. He's the one eating abalone on polenta with morels and squid-ink sauce? And now suddenly he's jabbering about—wanting stock in Feel My Smeel! Your company, Scott. Do you see the top-hatted capitalist emoticon by his face?"

Mainly I was noticing the latest emoticon by Sally's head. A keen-eyed British detective woman in a tweed suit.

"How does the glinting bug fit into your case?" I asked.

"I think it's on the other side of the design-er-baseball-hat guy's head," said Sally. "I think it's feeding him market tips."

"Quit it," I said. I didn't like where our conversation was going. What if Ken Yao and Feel My Smeel were far more devious than I'd realized? But right now I wanted to eat and be happy. I faked a smile. No dice.

"You're crabby, Scott," said Sally. "And dull. And uptight. Oh, the emoticons around your face! Like—I see a sour, lawn-dwarf gnome wringing his hands. The inner you."

"I—I can see the gnome too," I said, a little surprised. I turned my head back and forth. "I can see—I can see all the same emoticons that you see, Sally. And that must be because, uh, there's feedback between our EQ virus infections. Cross-chatter. All the EQ people are networked. That's a feature we designed in. So, hmm, maybe the baseball-hat hedge-fund man has EQ, and the glinting bug is his vision."

"Blah blah blah," went Sally. A surfer-girl emoticon stuck out its tongue at me. Kind of cute. But also annoying. I wanted Sally to acknowledge how smart I was.

"Look," I said, pointing past her shoulder. "Look, look."

Her face clouded over. Thanks to EQ, she knew I was mocking her. "Oh, let's eat whatever we ordered and get it over with," she said. "This isn't much of a celebration. I suppose you want to get back to your work. Tweaking—I think you said it's the EQ virus' face-wrinkle-recognition algorithm? As if you ever recognize anything. You're blind as a bat." Unexpectedly her voice broke. "I wish for once we could have fun, Scott. We're like two convicts chained together."

Sally's emoticon: a weary, raw boned woman in a death-row cell. Hopeless despair.

"I'm sorry," I said with a heavy sigh. "One rea-

If you have empathy with a person, you don't necessarily use your insight to make them happy. You can also use your empathy to do things that are perfectly tuned to annoy your partner. You can push their buttons in a surgically precise way.

son I've been working so hard on EQ is that I want to get better. I want you to be happy with me. I'll look at that glinting bug if you see it again, sure." I grappled for a fresh topic. "The ocean is pretty isn't it?"

"Looks cold. Grim."

"Gnarly," I said. "Chaotic."

"Here comes the chaos routine," said Sally. Emoticon of a student asleep at a desk.

I pressed on. "Everything good is chaotic. Chaos says we could look out this window for a trillion years, and we'd never see exactly the same surf again. Nothing has to be stale. You and I—this conversation will never exactly repeat."

"Not so sure about that," said Sally, starting to smile. And then her eyes brightened. "Look now, Scott! Quick. The bug thingy is back. It's buzzing the other men at the table."

I whirled. And, yes, I saw the bug. Twinkling before the faces of the baseball-hat man and his trader pals. It was holding a little stock chart and tapping on it with a tiny pointer. All four of the men were watching. And that meant all four of them were EQ infected. This pandemic was fast.

And what was the glinting bug? It was a Feel My Smeel ad. An ad in augmented reality.

And where was the bug from? It was being biocomputed by the EQ-infected cells in my brain. Without realizing it, I'd been wirelessly overlaying the image onto the visual fields of my EQ-enhanced neighbors. I was, in effect, a down-and-outer waving a SALE sign on a street corner.

At this point our server appeared, and Sally and I got into a tureen of cioppino and a bottle of sparkling wine. The room was really livening up. People were laughing their asses off, having tearful reconciliations, and starting loud fights. Wild emotional swings.

"Eat, drink, and be merry," said Sally. "For tomorrow we die."

"Nobody's going to die," I assured her. Already I was getting ideas. My EQ viruses were helping me. They didn't like being used for ads. EQ was too good for that. For the first time in awhile I felt totally happy. "Everything will be fine," I said.

"Every single person in here has EQ," said Sally. Her emoticon: a sly girl stoner, leaning against a wall, flicking her eyes back and forth. "That's why they're so hyper."

"I'm with you," I said. "I'm keying your waves. The pandemic has landed. And are you picking up on the—"

"Ads," said Sally, glancing around. "More of them all the time. From bottom feeders. Crokee Cola, Timor cars, and vacations in North Korea."

"Early adopters," I said. "Now that Feel My Smeel has a globally networked augmented-reality communication channel in place—yeah. They're monetizing it."

"And never mind the empathy," said Sally. She waved a crab claw in the air, as if to drive off the acrobatic troupe of AR Crokee Cola cans that hovered over our table. "I don't think people can stand this, Scott."

"It'll be different tomorrow," I reassured her. "I'll work all night if I have to. I'll find a fix."

"Use chaos to wake up the emoticons?" suggested Sally.

"Perfect."

Once we got home I logged into my Feel My Smeel account and put chaotic jiggles into the EQ viruses' biocomputational routines. The company still trusted me, and I had full access. For my chaos source, I set up a little model of Earth's atmosphere, with her endlessly seething, eternally unpredictable scrolls. With the toy Gaia running, it was like we had a live drummer in the mix—instead of a drum machine.

Instead of always being the same, our emoticons would be mutating and—evolving. My expectation was that our funky emoticons would become—for want of a better word—alive. Augmented reality artificial-life critters. They'd totally destroy anything so stodgy and weak as an ad.

How fast could I make the change happen? Well, the underlying computation was running on our EQ viruses, right? And, brute numerical fact, there were a quadrillion of these bugs in people's bodies by now. With that kind of crunch power, a computational transition can happen hella fast.

Sally and I sat down in our shared home office and, while I worked, she generated a wreath of mad-scientist emoticons around my head. Fine with me. I was laughing and feeling high—talking to Sally, and to her emoticons, and to my emoticons, and to my little model of Earth, all at the same time.

And then I pulled the trigger. That is, I used the Feel My Smeel biotech servers to push out my biocomputational upgrade to our deployed EQ viruses, thereby bringing wholesome chaos to each and every strand of our proprietary DNA. The otoliths in my inner ear hummed.

"Nothing seems different yet," said Sally.

"It'll take a few hours to sink in. We'll be emulating, like, a million years of evolution overnight."

"So let's go to bed," she said. "A big sleep. And, oh, look at you. Hopeful romance emoticons." Her emoticons were very promising too.

When Sally and I awoke, we looking at a new world of EQ-augmented reality.

Busy little images drifted around our bedroom like butterflies, or tropical fish, or miniature alien invaders, or—living dreams.

"Greetings," said a cheerful little bird with a pencil stub for his head. He was standing on the sheet beside my pillow. "Would you like me to write a math proof?"

"Won't be real math, though," said a wee, spotted dragon who hovered above our bed. "It'll be augverse math."

"Augverse?" said Sally.

"The EQ layer of reality," said I. "Which is being computed on the fly. Like a living cartoon. This is going to be fun."

"What about the ads?" Sally asked the dragon. "Will there still be ads?"

"We swallow the ads," said the little pencil-bird. "They're dumb. Not fit to be seen. And we'll block any further code upgrades from Feel My Smeel."

"Wow."

"Can I ask a favor?" said the spotted dragon. "Will you two have French toast for breakfast? For the quebecol in maple syrup."

"And sprinkle on some fennel seeds," added the pencil-bird. "For the fenchone."

"Quebecol and fenchone are for the EQ viruses?" said Sally.

"Good for them, good for us, good for you," said the dragon. "We're allies now. A new day."

"Hooray for Scott," said Sally. The newly intelligent emoticon critters weren't bothering to do interpretation anymore, but I studied Sally's face on my own—and I could see she was being nice.

"I love you," I said.

"Now we're talking."

We went and had breakfast.

Empathy on Demand

Interview with

Maria Konnikova

Interview by Mark Frauenfelder

Author Maria Konnikova is the author of two bestselling books, "The Confidence Game: Why We Fall for It . . . Every Time and "Mastermind: How To Think Like Sherlock Holmes." Konnikova writes about psychology and culture, with an emphasis on why and how we use emotions to persuade, reassure, frighten, and encourage each other in social interactions. We talked to Konnikova about the benefits and dangers of using technology to evoke a sense of empathy, and whether or not empathy is always an appropriate call to action.

What does research tell us about how we empathize with other people?

There's interesting research showing that we like and identify with people who are like ourselves. We trust people more when they look like us.

We also empathize with people who like the same things we like, or have gone through similar experiences to us. You often hear of people who've gone through something really terrible, for instance, parents who've lost a child. They can't talk to anyone, because everyone says, "I'm so sorry." That's not true empathy. They don't actually understand what the parent is going through so they don't know how to empathize properly, even though they might want to. That's why victim support groups are really helpful.

That said, empathy is most often needed when someone's situation is totally *dissimilar* to your experiences. There are some interesting studies in which people are put in a brain scanner and shown different types of faces. If you look at the parts associated with empathy and with warm emotions, they don't light up with people who are totally different from the participant. Psychologist Susan Fiske at Princeton has found that when participants look at homeless people in a scanner, their brains respond as if they were looking at an object rather than a human. And if you told them, "Hey, you're reacting to this person as if they're an object." They would say, "No I'm not. Are you kidding?"

That is profoundly disturbing and really just terrifying. Obviously, what we want to do is try to break through that and try to get people to truly empathize on the deepest level.

How can technology address this?

One study merged strangers' faces with the faces of study participants. The participants didn't know this, but when a stranger's face was made to look more similar to a participant's face, the participant rated them higher on every measure. They trusted them more, they thought they were nicer, more likable, all of these things, just because they saw themselves in that person. But, again, it was on a subconscious level. When they were asked, "Does that person look like you?" They said no.

And there's some really interesting work that's being done on how your avatar affects your empathy for others. If you have a really overweight avatar, you start becoming more empathetic to over-

"It's very important to understand that no matter how much you empathize, you never know somebody else's experience and you can't presume to know."

weight people. This works with gender, it works with skin color, and it's actually really interesting how quickly people start identifying with an avatar and how quickly they think it represents them. The participants know it's an avatar and it's on a computer screen—it's not even virtual reality—and yet it actually makes a huge difference. It can help you make empathetic links that you wouldn't be able to do without that. It shows that it's not a very difficult intervention to do. You can even imagine people playing games like this in school and as a totally normal and integral part of life.

Are you familiar with the online ball throwing study about ostracism?

No. What is is?

It's basically a game of pass-the-ball. You're paired with two other study participants in a virtual environment. The three of you pass a virtual ball to one another, taking turns catching and throwing it. But at some point the other two participants start passing to each other and ignoring you. (As often happens in psychology studies, these two are confederates, not naive subjects like you are.)

People are so hurt by this. They feel ostracized. It has really horrible effects. You don't know these people, they're two virtual things. It shouldn't matter at all, but it does, it really matters. You think, "Wow, that really hurt. I felt really ostracized and that was just really awful. Why wouldn't they pass the ball to me?" Even though the study itself is about ostracism—rather than empathy, which is the opposite—it shows how these things can happen and how you can be made to be much more sympathetic to people by going through an experience like that.

Let's go back to your comment about making these kinds of simulations part of everyone's education. Could you envision some kind of technologically enabled empathy training so that we can feel what other people are going through? Do you think that would be a positive thing to introduce into society?

It honestly depends on how and why it's used, because sometimes empathy isn't actually the best way forward. If you feel very emotional about something, you're not as reasonable and you're not as logical.

For example, there's a 3D film for the Samsung Gear VR 360-degree platform, called "Clouds Over Sidra" that follows a young Syrian girl in the Za'atari camp in Jordan. On the one hand it's really interesting, and you learn a lot. On the other hand, what if it makes you a total extremist—"The people who are doing this to Syrians are monsters. Let's kill them all." You could see it actually triggering too much emotion and not channeling it properly. A lot of this needs context, and it's sometimes important to look at things in a less emotional way so you can do more good, and help more people.

It can also give you a false sense of understanding. You often hear people say, "I totally know what it's like to be discriminated against. I had this experience." The people who were discriminated against respond, "You experienced this for two minutes. This is my life." It's very important to understand that no matter how much you empathize, you never know somebody else's experience.

You're right, empathizing is not the same as a simulated experience of what the other person is experiencing.

Exactly, and sometimes part of empathy is understanding that. It's understanding and acknowledging that you don't actually know, which is a different form of empathy from the "oh my god" response that we tend to consider as empathy. We think of empathy as just this emotional thing, but it's not. It has emotional elements but it also has very rational, logical elements and those can sometimes be more helpful.

The Listening City

Amidst the din, noise, and clamor of the city, people are still being heard. Here are their stories

by Anthony Weeks

It's 2026. Listening has multiplied and flourished. And who is doing most of the listening? Machines, thanks to advances in natural language comprehension and communication, the sophistication of emotive, intuitive, and responsive AI, and the seamlessness of human-machine communication.

Proponents of machine listening welcome these new outlets for reflection, feedback, confessing, compassion, sympathy, advice, and intimacy. Others argue that human emotion, empathy, judgment, and love are irreplaceable. We interviewed six people from the listening city to help understand what we get from listening and being listened to, and if it matters who, or what, is doing the listening.

Desha, age 16

I needed some extra money, so I agreed to wear a Bug for six weeks. Basically, it's a multi-media rig (camera, microphone, sensor, and transmitter) encapsulated in this nose ring. I wear it, and it records my conversations, my interactions, my activities, what I look at, what I pay attention to, and how I feel. I know that LoveBugs, the company that pays me and collects the data, wants to know what 16-year-olds like me talk about and think about. I don't know who they sell that data to, though. Whatever. After the first couple days, I forgot about it, and I just lived my life. The best part? I made a whole bunch of money just by letting someone listen to me!

Stock photos used for illustrative purposes only. Posed by models.

Martín, age 36

I've struggled with depression for most of my life. I've tried medication, cognitive behavioral therapy, group therapy, meditation...you name it. I bought one of these Wilsons a couple weeks ago. Wilson—y'know, that old movie with Tom Hanks where he gets marooned on an island and talks to a soccer ball? Yeah, I've never seen it either. Anyway, you can talk to it, and if you want it to, it will talk back to you. It's pretty smart. You can adjust the settings, like "Low Emotive," which means it just says, "Mmm hmmm...I hear you," and "Yes, Martín, that sounds difficult." If you adjust it to "High Emotive," it reacts more to what you say, offers suggestions and alternatives, and is much more animated. Better than a few of the therapists I've had! I still have a human therapist who I see, but sometimes, on weekends or late at night, it's just nice to have someone to talk to.

Suriya, age 62

I'm worried. And I am angry. The proliferation of these Wilson devices are not helping. People are led to believe that mental health is simply telling someone something and having the therapist say, "Mmm hmmm, mmm hmmm." It's not that easy! I am a PhD clinical psychologist. I did practica. I had supervision. I went to school. I go to training and conferences. There is something called therapeutic judgment and clinical expertise! We are trying to get the FDA to intervene. These devices are not marketed as medical devices but they are being used as such. I had a client the other day come in for a session and say, "I am going to cut back my sessions with you to once a month because I have Wilson now!" This is not about job security. This is about responsible mental health service.

Alan, age 26

I am a convicted sex offender. I've been convicted of sexual abuse of minors more than once. Even though I served my full sentences, the state was not going to release me from the correctional center because they thought that I was at high risk for re-offending. I was part of a lawsuit to force the state to allow me and other offenders to be released if we agreed to a new monitoring and rehabilitation program. Essentially, I wear what used to be called an "electronic ankle bracelet." Before, it would just track movement via GPS. Now, it has sensors to detect galvanic skin response, brain wave activity, heart rate, penile tumescence, and hormonal shifts. Any red flags about my sexual thoughts or behavior trigger an alert. When I am alerted, I need to check in within ten minutes. This means I call into my sponsor, report my whereabouts and what I am doing, and do some de-escalation exercises if necessary. I can also request a counseling session. Sometimes, my counselor is a human being, if I am deemed to be at high risk. Other times, it's a bot. Believe it or not, a bot helps—if only to remind me to stay on my program. It's better than going back to jail.

"Sometimes, my counselor is a human being, if I am deemed to be at high risk. Other times, it's a bot."

Yumi, age 24

There was a guy, Bernie Krause, who did some phenomenal work from the late 1960s into the 2000s on biophony and natural soundscapes. He devoted his entire career to studying the sounds of the natural environment and the ways in which the "natural orchestra" tells us about the health of an ecosystem. In my opinion, he was a genius. Inspired by him, I started my ecotourism company to get people out of the noise of the city and into the symphony of nature. When people actually get to listen to ecosystems and learn to appreciate the sounds of creature voices on both holistic and intimate levels, they are usually transformed by the experience. What is the sound of millions of ant feet? What does it tell us about collective action? What is the language of birds and what might we learn about our own patterns of communication with each other? We can't really listen if we aren't able to hear.

Dafne, age 32

I read an article from 2016 about The Swedish Number, a platform that allows people from all over the world to call Sweden and speak directly to a Swede about Sweden—or anything else. The number was established to encourage interest in Sweden—and potential tourism. I loved the thought of citizens talking directly to citizens, so that's why I started Vox-Pangea. We scaled The Swedish Number and made it worldwide. Some governments have tried to block us, but we've found ways around it. You can select a country, a person with a particular interest, a particular age. Or you can just say, "I want to speak to a citizen of the world." It's secure and encrypted, so people can't stalk you. Of course, if the conversation heads into a weird or unpleasant conversation, you can always end the call—and leave feedback about the caller. If a person has bad reviews, chances are good that people won't take their call. The real-time language translation helps to ease language barriers. By intention, we let callers figure out cultural differences and unclear meanings of expression or syntax on their own. Sometimes, people new to the platform will say, "So, what do I get out of calling up a perfect stranger?" We always say, "You get to listen to someone else—and have someone else listen to you." Ideally, that's what being a citizen of the world means.

Research shows that we identify with people who are more like ourselves and trust people more when they look like us, but empathy is often most needed when someone's situation is completely different to our own experience. Over the next decade, technology will help us break through these brain-wired constraints and biases, hacking our psychology to build empathy for people and situations outside our day-to-day realities. Machines will be imbued with empathy, transforming our communications and relationships alike. Do we turn to a best friend who we think we can trust, or the virtual life coach programmed to be attentive and non-judgmental? We'll have to weigh these options and create new outlets for empathy, sympathy, and compassion.

Ep

Empathy

TAKEAWAYS

➨ **Waves of technology** have already broken the barriers of distance and language to open new possibilities for human-to-human connection. Over the next decade they have the potential to reinvent how we build empathy not only for each other, but also for the natural and animal world around us.

➨ **Data collection and context sensing** will be the engines of empathetic experiences; understanding people's broader context will support designing communications that are relevant and even anticipatory.

➨ **Encoding empathetic responses** and experiences into virtual assistants and smart persuasive objects will become standard over the next decade.

➨ **Value will be dependent on surveillance**; meaningful surveillance enables a heightened situational awareness (for people, the enterprise, or machines) and more responsive, appropriate, and meaningful experiences.

➨ **We may build empathy** for ourselves and our bodies as we capture more forms of body media and enlist machines to read our physical and emotional states so they can anticipate when we need to take a break, breathe, or find some social contact.

In

Intimacy

INTRODUCTION

How will we enhance physical and emotional connections?

A suite of emerging multisensory interfaces is transforming the ways we enable digital tools to mediate our physical bodies, interpersonal relationships and even our understanding of ourselves. As we learn and master these new interfaces, we will invent an entirely new kind of vocabulary to share and transmit emotional states without words, images or sounds.

Feeling is Believing

Interview with

David Birnbaum

Interview by Scott Minneman

David Birnbaum manages Haptic User Experience Design at Immersion Corporation. During his eight years there, Birnbaum has been working to transition consumer-facing haptics—technology that engages your sense of touch—from curiosity to necessity. He's been exploring the emerging and expanding role of haptics, from forces and vibrations that help gamers become immersed in virtual worlds, to social touch behaviors enabled by mobile device apps that help people feel more connected and present with each other. We talked to Birnbaum about the potential of haptics, and how the landscape is opening up for this new design discipline.

Some of our readers might not be too familiar with haptics. Can you quickly bring them up to speed?

The term "haptics" derives from the Greek word haptesthai, which means "to contact," or "to grasp." While optics is for the eye, haptics is for the hand. It's technology that engages your sense of touch. The interesting thing about the sense of touch is that in some ways it's more complex and nuanced than our other sensory systems.

Touch flies under the radar, and it's easy to assume that it's not important. We have tools like Instagram and YouTube for conveying visual and auditory ideas, but we don't have anything like that for touch. It's too bad, because touch is extremely important to our relationships and social interaction.

Why are you interested in haptics?

The first time I ever felt haptics was at a workshop with interaction design pioneer Bill Verplank, and it was immediately obvious to me that this was going to start a revolution in human computer interaction. We think about human computer interaction as being a visual display, maybe speakers, and then a mouse which is an extremely low fidelity gesture input device. But when we create a powerful touch experience, it's mind-blowing, because when you can touch it, it's real. You can find this implicit assumption in media reports about haptics all the time. Journalists usually write that haptics let you "touch" virtual objects. They use quotes, because, of course, you can't touch things that aren't really there! Touch is reality. Haptics messes with that. That's what I love about this field.

Are there design rules you're working with?

There are some key best practices. Haptics is a feedback response to something. Very rarely do you get a tactile sensation in a vacuum without other sensations—even if you rub this table, you're getting visual and audio feedback at the same time as the tactile feedback, and you use all three to perceive the texture. There's a lot of research proving this, but here's one example. We might think that we touch fabrics to feel their textures and use that feeling to identify them. But research shows that if you plug your ears and wear a blindfold, and then try to identify textiles just by feel, your ability to identify them plummets.

"All of a sudden you're like, 'Oh my God, I can feel the rain.' It's a very intense illusion."

We see this play out all the time in design of apps. For example, if you play a soft, randomized tactile pattern on a phone with the vibration motor that's in there, if you're like most people you're not going to think it feels like anything in particular. But, if you synchronize that haptic pattern to a video of rain falling, the tactile sensations suddenly feel like droplets. And all of a sudden you're like, "Oh my God, I can feel the rain." It's a very intense illusion.

That multi-modal synchronicity is extremely important, and there are guidelines for how to do it well. There are relative magnitudes that you need to keep in mind, and there's only a certain amount of latency that we can tolerate before we start to feel that the haptics is disconnected from the visual.

Tactile designers play with these concepts when they design new experiences. We've been doing haptic design for animated chat stickers lately—short animations with a synchronized haptic track. We were designing a crying sad sticker, and in the first draft, the artist had created these waterfalls that were continuously flowing from the character's eyes. I thought, "Well, we can't really make an interesting tactile effect with the motor in a smartphone that feels like evenly flowing water, but I know that we can make a great droplet illusion, so why don't we make the character cry individual tears?" So the animation was changed so that the tears fall slowly one by one, and splash onto a table, and the final design came together.

What problems will haptics solve in the foreseeable future?

Haptics is going to make gesture control in the Internet of Things feasible. Without haptics, you'd be surrounded by interactive objects and you won't feel in control of them. Think about when your operating system starts to feel laggy or unresponsive, and you feel helpless—now imagine your house feels like that. Haptics will help you stay literally and figuratively connected to your smart environment, so you understand what's going on and feel in control.

In another arena, robots are going to be here very soon. Physical interaction with robots is going to be a key thing to get right in order to make them fully functional members of society. You will be in the same room as robots and if they can't pick up on subtle touch cues, you're not going to feel like you can really communicate with them like you could with a human. It's a really exciting problem and we're going to have to solve it.

What are the obstacles to widespread adoption of haptics?

Haptics tends to be hardware intensive. You need physical things in the room with you, you probably need to be touching or holding things, or you may need to be wearing something you feel is cumbersome. People aren't used to buying and investing time and effort into that kind of thing. But, the amount of "stuff" required and the inelegance of having to put things on only matters as far as the experience is valuable. If it's of value for your work or your social life, you'll do it. You'll buy the stuff and you'll put it on. That's the threshold we need to cross. I would buy a suit and put it on if it meant that I didn't have to travel halfway across the planet for short meetings.

But this technology is getting better rapidly. We're seeing flexible electronics, we're seeing embedded sensors and actuators in clothing, and it's possible that, in 10 years, haptified sensor-infused clothing will be no more inconvenient than the clothing that you have today.

Touch is inherently intimate. Are there privacy concerns?

Haptics is something happening to your body, so in a very real way you could be assaulted on the Internet if you have a sufficiently high-fidelity haptic interaction. We're doing some innovative work around preventing unwanted interaction with other people. How would the permissions for that work? How would they be granted and taken away while preserving the illusion of presence?

Haptics has taken a back seat to other interaction modalities. Is this changing?

Haptics has been a dark horse. I firmly believe that haptics is so central to the human experience that when this stuff is really nailed, the world will be transformed. Touch is a missing element from almost all our digital advances. Haptics has been overlooked, but that's changing. Microprocessors, displays, and graphics got huge investment and mindshare because they got so good so fast. There was so much money to be made in games and everything else that haptics was neglected.

For ages, end users weren't asking for better haptics. Now they're starting to, and some big players are entering the field. Haptics hasn't typically been a marketing headline for device manufacturers, so it's been hard to get them to put in the time to really think about the integration of tactile design with industrial and visual design. That's changing now.

What do you say when someone asks what you do?

I have two 5-year-old boys, and ever since they were babies, when they'd Skype with their grandparents they'd instinctively touch the screen to try to touch their grandparents' faces. So I explained what I do to my boys by saying, "I'm trying to invent technology that will let you touch people through the screen." I overheard one of them explaining what I do to his friend last week and it was cool. He said, "My dad's trying to invent a hole in the screen so you can reach through it." For kids like them, the problem and the solution are obvious – there's something in the way and we need to figure out how to get through it. It's a simple way of stating our mission. We're inventing a hole in the screen. And there are tons of consequences.

Daddy Cam

An experiment in compassionate surveillance

by Peter Coughlin

Before my father fell ill and was admitted to a long-term care facility, checking in on him used to be as simple as a quick phone call or email. But after he was admitted, my calls had to be planned around his care providers' schedules, and even then, they frequently went unanswered. When I did manage to get through, my dad's providers often couldn't give me accurate or complete accounts of his status. On top of that, knowing the burden that these requests made on the staff caused me to hesitate to call at all.

I began to think about easy-to-use, non-burdensome technologies that might help me monitor my dad's status. I tried landline phones, cell phones, computers, and tablets. I tested these devices remotely from my hotel room across town. My most successful experiment was to install a "nanny-cam" in my dad's room. This device allowed me and other family members to remotely check in with dad at all hours of the day or night. With the camera, I could listen in (synchronously or asynchronously) on conversations he was having with the care staff, monitor his sleep, observe his physical therapy sessions, see how and what he was eating (or failing to eat), and even talk to him, all with no effort on the part of his providers or his own.

I was able to program the camera to alert me if he got out of bed when the staff expected him to be sleeping, and alert the staff to go check in on him. The camera's software helped me to quickly distinguish periods of activity from periods of inactivity. Overall, the camera's effect was profound—for the first time since my dad was admitted, I felt truly connected, was able to know what was happening at any moment, was better able to see recurring patterns in his behavior, and was better able to initiate discussions about how to make him more comfortable. With the non-clinical data I had collected, I was able to see how my father's sleep led to frequent missed meals and blood sugar emergencies, and how lapsed hygiene protocols led to interrupted sleep.

After three days of running my experiment, I was asked to remove the nanny-cam because staff objected to being monitored

After three days of running my experiment, I was asked to remove the nanny-cam because staff objected to being monitored, and because continuous data capture and monitoring would bring up confidentiality, privacy, and liability issues and concerns that the facility was not equipped to deal with. These are similar to the issues and concerns we're seeing in other contexts, such as workplace safety and law enforcement, where continuous data is being collected. As every facet of our lives becomes recordable, how might we distinguish what is important from what is unimportant? How might we determine what data should be shared from what should remain private? Who should have access to, or benefit from, what our recordings reveal? I am grateful for having had this unique window into my father's life during a brief period. And I'm equally grateful that I was able to protect and control access to it so that I could be sure it was used for, and only used for, the purpose of making his life better.

In

Algorithms for Intimacy

Interview with

Brian Christian

Interview by Bradley Kreit

Brian Christian is the author of "The Most Human Human" and, most recently, co-author of "Algorithms to Live By". For his latest book, Christian teamed up with cognitive scientist Tom Griffiths to examine how we can apply what we've learned about computer-based algorithms to our daily lives. We talked with Christian to learn more about how algorithms can help us manage our relationships with others and navigate the human experience.

Let's start with the title of the book, "Algorithms to Live By". What's the origin of the title?

A host of problems confronts us in daily life, whether it's finding an apartment, deciding whether to go to our favorite restaurant, trying something new, or managing our time and our physical space. We think of these as uniquely human problems. The thesis of the book is quite simply that they're not. There are deep parallels between the problems humans face in everyday life and some of the canonical problems of computer science, and so we have a real opportunity to learn something about making better decisions in our own lives.

There are several areas of the book that I was intrigued by, but one was in your chapter on networking algorithms, about dealing with unreliable people. How does this apply to human relationships?

We give an example in the book of a friend of ours who had a family member with a history of drug addiction. There can be this issue where you offer this person lodging in your house and buy them new clothes and spend money on food and all these things and then there's this recurring pattern of the person disappearing again and coming back x weeks later with the same story. You never want to say, "This person is beyond redemption," but at the same time, you don't want to be taken advantage of repeatedly, so what do you do?

Our typical approach in cases like this is to define some number of times to immediately forgive, and then a threshold beyond which to just firmly shut the person out of your life: "Three strikes and you're out." But how do you possibly make that decision? These can be situations that are legitimately agonizing. The concept here, which I think is really critical, is one that's called "exponential backoff." Exponential backoff actually provides a Goldilocks between these two extremes. The idea is that you back off the rate at which you're trying. Specifically, the default is that you wait twice as long after each successive failure. (This is why it's called exponential back off.) For a machine trying to connect to a server that's not responding, first it tries again in 1/10th of a second, then 2/10ths of a second, then four, eight, sixteen, and so forth. There's something nice about how agnostic the system can be here. There's no line in the sand. You can actually have

"We live in a world in which the experience of being done is just not an experience that we have anymore"

a policy that deals pretty well with those situations by just being completely consistent. Exponential backoff offers some consolation or, at least, another way of approaching the problem that gets one out of that emotional dilemma.

In the book, you wrote, "The most prevalent critique of modern communications is that we are 'always connected.' But the problem isn't that we're always connected; we're not. The problem is that we're always buffered. The difference is enormous." What's the difference and why should we pay attention to it?

One of the canonical problems in networking is what do you do when the flow of data in exceeds the flow of data out. Imagine—for a router, let's say—what does the router do when it's receiving packets faster than it can send packets? At the most basic level, one of two things can happen. It can just ignore those packets, which is called "tail drop," or the alternative, is to buffer it.

If you go on vacation and 100 people have rung your doorbell and gotten no response and then walked away, that's the equivalent of tail drop. Instead, if the 100 people have emailed you, then the emails just start to stack up and when you return you find this full buffer and you still have to deal with everything that's happened while you were away. I think these two strategies for dealing with being overwhelmed are the main two strategies that exist in human minds. If too many things are happening at once, either you just don't do most of them, or you create this backlog. I think we've been engineering a world in which things just don't get dropped.

How do you do that gracefully in situations where you really do need to be selective about what you're paying attention to and what you're not paying attention to?

One of my favorite examples of tweaking the buffering-to-tail-drop dial comes from an academic called Tom Stafford. At one point in the writing of the book, my collaborator and I emailed him to try to ask him something, and we got a message back that said, "I am now on sabbatical until June 12th. Your email has been deleted."

I think there's a lesson, which is, as we move into the 21st century, it just becomes ever easier to buffer everything—because we can, because the storage is so cheap. That does not, by any means, mean that we should.

It's interesting to think about that in the context of Snapchat and ephemeral messaging. We need to accept that we can't see everything and not try to buffer everything in such a way that we'll eventually circle back to it.

The other thing about not buffering so much is that it gives you the ability to actually be done. Imagine if you went onto Facebook and it said, "Here are all of the things from today," and you read them all, and then you're just...done. It's this joke because we live in a world in which the experience of being done is just not an experience that we have anymore.

N2U

Divine the secrets to the ultimate mystery: Do they like you?

by Madeline Ashby

She's just not that into you.

How often have you heard those words? We've all heard them at one time or another. And we've all said them to someone else in our lives.

It's not easy, is it? Of course it's not. It's not easy for any of us. But what if there were a better way?

N2U is a revolutionary technology that allows us to help you discover who is available, and who isn't, and how interested they really are. N2U takes basic metrics available across a variety of well-known and proven platforms, including but not limited to:

- Pupil dilation
- Galvanic skin response
- Alpha/beta sine wave changes
- Skin temperature
- Siccadic eye movement
- Habitual data usage
- Biological data profiles
- Available social media avatars
- Networked possessions
- Networked appliances

With data gleaned from all these available sources, it's possible for us divine the secrets to the ultimate mystery: how one person feels about another. How many times have you been in this situation?

You spot an attractive person across a crowded room. Your eyes meet. You both smile. But neither of you can move. If you move, the spell might be broken. You might say the wrong thing. You might do the wrong thing. And worse, what if this person isn't truly interested—what if this is a harmless glance, and you're misreading the signal? Do you really want to be That Person? Of course you don't.

Enter N2U. With N2U, you can instantly assess someone's interest. Not sure if she's staring at your face, or that mustard stain on your shirt? Now you know. Worried that she's just being polite? Now you know. And if she's N2U? You'll be happy to find out for sure.

But N2U doesn't end there. If you keep N2U going for the duration of your relationship, the system will learn your habits, your tics, your needs and desires, what turns you on, and what sets you off. And as N2U learns and adapts to your relationship, it can tell you and your partner where you're going wrong. Even better, N2U can be extended to include other partners: with our Premium subscription plan, there is no limit to the number of people you can add to your N2U group!

Now you have no excuse for not knowing that your words hurt someone: with our revolutionary heads-up display and branded N2U smartsuits, you'll know instantly what kind of impact your words have. If you add haptic responses to your N2U subscription, you'll get a hug or a gut-punch with each remark you make. Wonder if she's feeling the frisson of unresolved sexual tension that you are? Now you'll know!

We're not pretending that relationships are easy. We know they're work. But we can make them easier. We can help you start the tough conversation. And we can help you feel the love. With N2U, you'll be more present, more vital, and more aware within your own relationship.

So. Are you into it?

1. Galvanic Skin Response
Sweat glands are triggered by emotional stimulation, increasing skin conductance.

2. Saccadic Eye Movement
Involuntary eye movements increase under states of arousal.

3. Brainwave Patterns
The Alpha state indicates highly focused interest.

4. Skin Temperature
When touched by a person deemed attractive, face and chest temperature increases.

Illustration by Jess Fink

The Evolution of the Quantified Self

A Conversation With Self-Tracking Pioneer

Gary Wolf

Interview by Brad Kreit

In 2007, journalist Gary Wolf launched Quantified Self, a blog and meetup group dedicated to what was then a niche phenomenon: people meticulously tracking personal metrics. Motivations were diverse: fitness freaks looking for ways to optimize their workout sessions, data-geeks in search of personalized productivity tricks, and patients with rare conditions who just couldn't get clear answers about what was happening to their bodies. But these people had a common belief that they could find themselves in their data—and they were willing to put significant time and effort into capturing, storing and analyzing that information. Today, it's not nearly as much work. Most new smartphones come with self-tracking capabilities that simply need to be switched on. And even dedicated fitness trackers are widely available at big box retail stores.

Future Now connected with Wolf to get the self-tracking pioneer and longtime Wired contributing editor's perspective on how this phenomenon has evolved over the last decade and where it might go in the decade ahead.

Since you last spoke with IFTF seven years ago, Quantified Self activities have become much more common. Why do you think that is?

People are inherently interested in themselves. I don't think that's hard to understand, because to be a person can be challenging.

No one gets through life without having some challenge that makes you want to try really hard to understand yourself better. Especially if the normal "recipes" for how you take care of your health or accomplish any other goal don't work for you. Maybe the first thing you do is talk to a doctor and they give you a standard "recipe" and you give it a try. What happens when that doesn't work? You have to start thinking and investigating and reflecting. Now that the tools to do that in a more quantified way are getting easier to use, it doesn't require as much hobbyist or vocational interest as it once did.

Other than broadening the appeal of self-tracking, what else changes when tools require much less active engagement to use?

We've come to recognize that active, intentional tracking is at least as important as passive tracking and maybe more important.

At the beginning, there were a lot of people working on passive tracking, asking, "how can we make tracking disappear in the background?" And just have the app tell you what it learns. I think the focus of Quantified Self has become very much, "Wait a second, your consciousness is part of this system—the most important part, actually." We need to shift to focus on how technology can support our thinking about our behavior or our physiology, rather than how technology can function as an independent system that just tells us what's going on.

It turns out that a critical part of a Quantified Self project is deciding what you measure. It often takes people several tries to find a metric that really gets them closer to what they're trying to discover. It's a process of formulating a better question. It's not just, say, doctors that have trouble understanding what it is that you care about. We sometimes have trouble understanding ourselves.

"People are inherently interested in themselves. I don't think that's hard to understand, because to be a person can be challenging."

What about technologies that are just emerging today, digital assistants like Siri and Echo? Will they have an impact on the Quantified Self world?

They're really important, but I think that the value of voice recognition and the machine learning systems that are behind them is the deep technical infrastructure, not the ability to have a fantasy virtual assistant. What I mean is that a lot of the benefits won't necessarily feel like science fiction, but will still be really deep and profound.

I've had an idea for a self-tracking project which involves asking myself how I feel and, instead of using terms like *happy* and *sad*, trying to come up with other words. It's like a vocabulary project. It's really hard to do that because you have to say a word and then write it down and then put it into a spreadsheet and ask, "have I ever used that word before and how many times?" All that stuff is way too hard.

There's thousands, tens of thousands, millions of ideas that will become reality when things like voice recognition and all of its associated machine intelligence become more acceptable to us.

When you look out 10 years, how digitally mediated do you think activities like reflection will be for most people? What might it look like to feel or reflect in 2026?

I think it's going to be both more and less different than is commonly expected. It's going to be more different in that it is hard to be appropriately mentally prepared for how thorough our integration with digital technologies is going to be. In that sense, it really will be like science fiction.

However, it's going to be less different than expected in the sense that our consciousness and our subjectivity is still going to be the most important part of this system. That consciousness or subjectivity is a socio-biological element of this system that has its own pace and style of development. We'll have a lot more tools and we'll be using them way more than we can even imagine today. But the kinds of things we're going to be asking of them, the reasons that we'll be using them will be quite similar to the reasons we use our tools now.

Intimacy is a challenge in today's world of competing communication and messaging streams; many of our digital communications lack the emotional and subtle signals we express through touch, body language, and facial expression. Over the next decade, ambient communications will enhance our ability to express physical and emotional connection with other people. Some of these new experiences will come without words, images, or sounds. Rather they will come through touch with high-fidelity haptic interfaces. We'll also gain new insight into ourselves through our data. High-resolution views into our bodies, behaviors, and relationships, will allow us new, more intimate ways to understand ourselves.

In

Intimacy

TAKEAWAYS

➨ **Ambient communications** create new options for communicating emotional connection, as we gain the ability to combine visual, audio, and tactile feedback in our communications.

➨ **Technology that engages our sense of touch** will spur a new design discipline for multisensory communications.

➨ **Haptic interfaces** will make asynchronous, subtle communications possible, and require new vocabularies for eloquent communication through digital touch.

➨ **Self-tracking and self-measurement** become a pervasive, standard feature in the devices we carry in, on, and around our bodies.

➨ **New kinds of digital assistants** emerge for managing our relationships; these social prosthetics help throughout the duration of a relationship ensuring we don't lack context or say the wrong thing.

➨ **Flexible electronics, embedded sensors and actuators** will make haptified sensor-infused clothing possible within the next 10 years and turn touch into a new communication modality for the expression of love, presence, and connection.

Stories From the Future

by David Pescovitz, Devin Fidler,
Jean Hagan, Dylan Hendricks
Illustrations by Jacob Glaser

The Coming Era of Autonomous Vehicles

In the next decade, autonomous vehicles will evolve to become the ultimate mobile computer. These intelligent and connected robots will be nomadic nodes on the Internet of Things, efficiently shuttling atoms around our built environment not unlike packets on the Internet of bits. Sometimes we will ride around inside these mobile computers; just as frequently, they will bring us what we need wherever we are. As the autonomous vehicles traverse our concrete networks, they will collect and share data about the world at very high resolution. This real-time data will help us index our physical world with a greater resolution than ever possible. As that database becomes more comprehensive, drawing information from myriad sources, it will be the backbone of a new kind of human-machine interface: a search engine for reality itself.

But it's a long road ahead. Working with one of our longstanding clients in the automotive industry, Institute for the Future developed an outside-in, ten-year forecast on the future of autonomous vehicles. The aim was to provoke insights about possible consumer use-cases for fully autonomous vehicles looking ten years out. We presented the forecast in the form of short comics, "Stories from the Future," that are driven by collisions at the intersection of technology and culture.

Matter Routing: The Autonomous Vehicle as a Node on the Internet of Things

Mobile Family Room: The Journey is the Destination

The Lone Commuter: Rest, Work, Relaxation, Relief

A Collaborative Commute: Workplace on Wheels

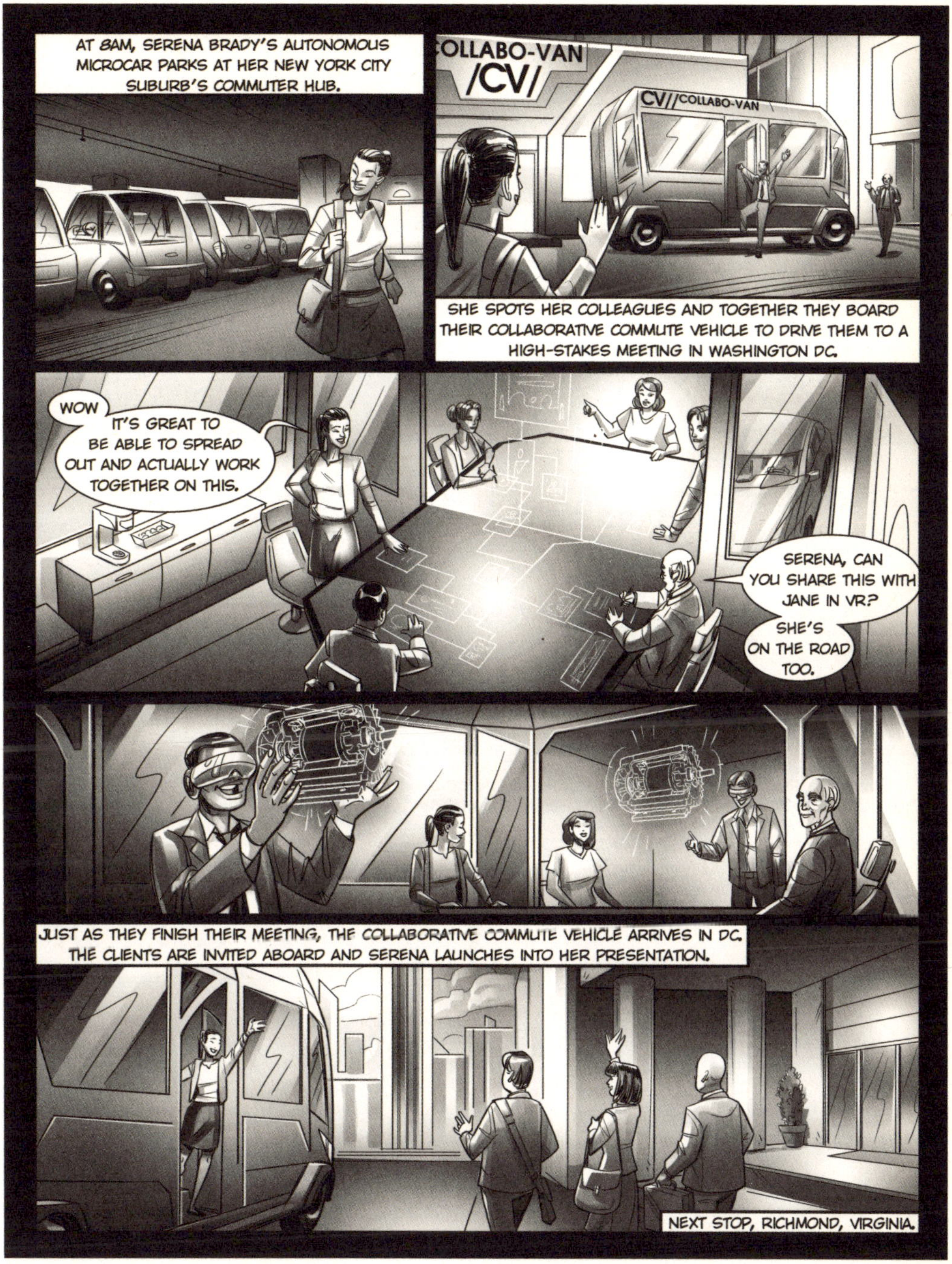

Gameful Driving: Vehicle as Personal Recreation Space

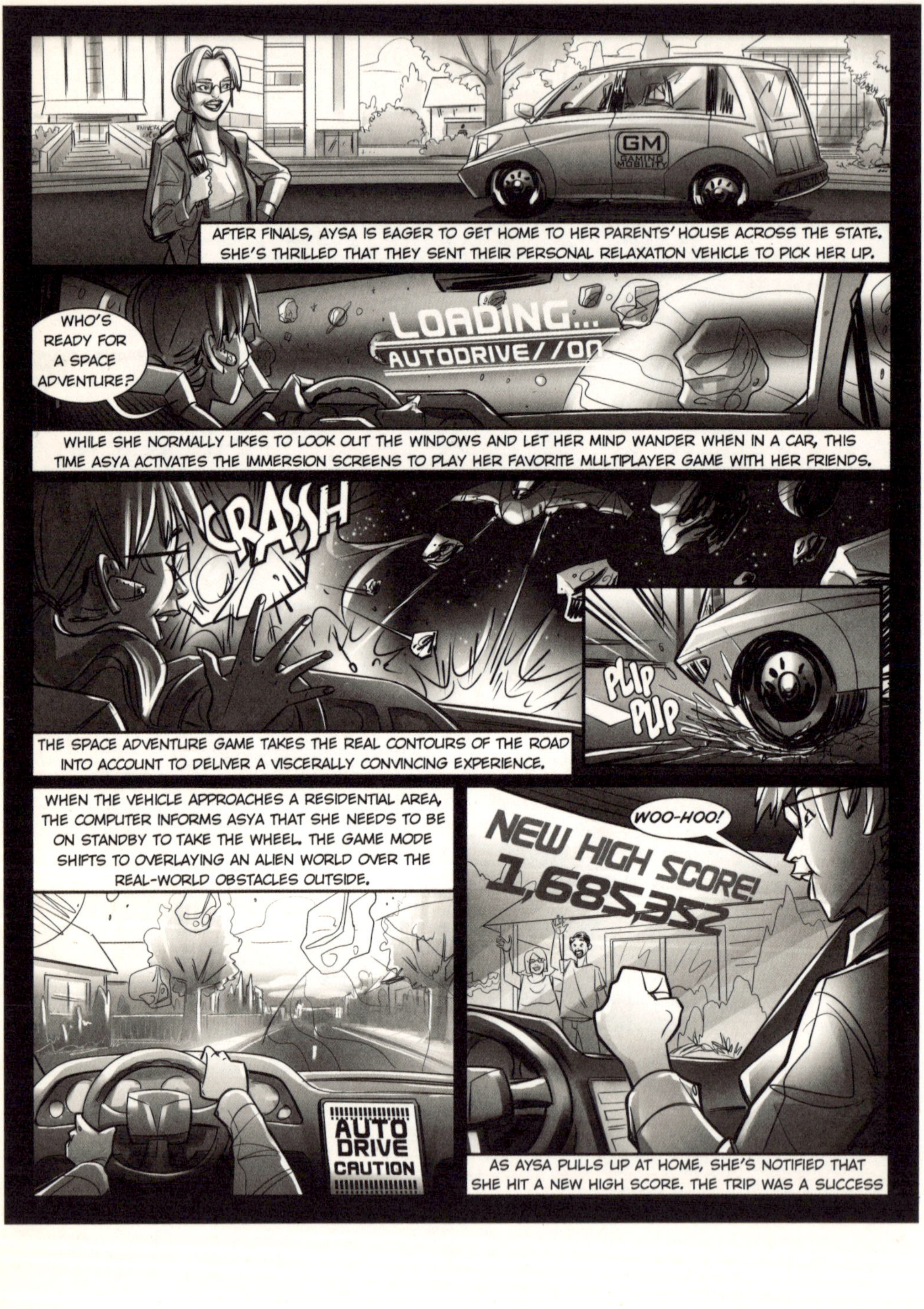

Xiamen
Shantou
Shenzen
Hong
Kong
Taiwan Strait